AF751044

Male Reproductive Anatomy

Edited by Wei Wu

Published in London, United Kingdom

IntechOpen

Supporting open minds since 2005

Male Reproductive Anatomy
http://dx.doi.org/10.5772/intechopen.94696
Edited by Wei Wu

Contributors
Yusuf Arikan, Ali Ayten, Tung Nguyen-Thanh, Thuan Dang-Cong, Oyovwi Mega Obukohwo, Nwangwa Eze Kingsley, Rotu Arientare Rume, Satoshi Hiroyoshi, Gadi V.P. Reddy, Hayfa H. Hassani, Fadhel M. Lafta, Rakad M. Kh AL-Jumaily, Maria Elizbeth Alvarez Sanchez, Oscar Rojas-Espinosa, Ereth Ameyatzi Robles Chávez, Maria Dolores Ponce Regalado, Edgar Antonio Estrella-Parra, Raul Aragón Franco, Jose Gadú Campos Saucedo, Cesar Lopez-Camarillo, Veronica Fernandez-Sanchez, Minerva Camacho-Nuez, Julio C. Torres-Romero, Abimibola Nanna, Victor Emojevwe, Osarugue Igiehon, Mohamad Eid Hammadeh, Houda Amor, Bakry Mohamed Sobhy, Almandouh Hussein Bosilah, Micu Romeo, Peter Michael Jankowski, Nyaz Shelko, Shridhar C. Ghagane, Rajendra B. Nerli, Makhadumsab Toragall, Murigendra B. Hiremath, Naiho Alexander Obidike

© The Editor(s) and the Author(s) 2022
The rights of the editor(s) and the author(s) have been asserted in accordance with the Copyright, Designs and Patents Act 1988. All rights to the book as a whole are reserved by INTECHOPEN LIMITED. The book as a whole (compilation) cannot be reproduced, distributed or used for commercial or non-commercial purposes without INTECHOPEN LIMITED's written permission. Enquiries concerning the use of the book should be directed to INTECHOPEN LIMITED rights and permissions department (permissions@intechopen.com).
Violations are liable to prosecution under the governing Copyright Law.

(cc) BY

Individual chapters of this publication are distributed under the terms of the Creative Commons Attribution 3.0 Unported License which permits commercial use, distribution and reproduction of the individual chapters, provided the original author(s) and source publication are appropriately acknowledged. If so indicated, certain images may not be included under the Creative Commons license. In such cases users will need to obtain permission from the license holder to reproduce the material. More details and guidelines concerning content reuse and adaptation can be found at http://www.intechopen.com/copyright-policy.html.

Notice
Statements and opinions expressed in the chapters are these of the individual contributors and not necessarily those of the editors or publisher. No responsibility is accepted for the accuracy of information contained in the published chapters. The publisher assumes no responsibility for any damage or injury to persons or property arising out of the use of any materials, instructions, methods or ideas contained in the book.

First published in London, United Kingdom, 2022 by IntechOpen
IntechOpen is the global imprint of INTECHOPEN LIMITED, registered in England and Wales, registration number: 11086078, 5 Princes Gate Court, London, SW7 2QJ, United Kingdom
Printed in Croatia

British Library Cataloguing-in-Publication Data
A catalogue record for this book is available from the British Library

Additional hard and PDF copies can be obtained from orders@intechopen.com

Male Reproductive Anatomy
Edited by Wei Wu
p. cm.
Print ISBN 978-1-83968-524-8
Online ISBN 978-1-83968-525-5
eBook (PDF) ISBN 978-1-83968-526-2

For EU product safety concerns:
IN TECH d.o.o., Prolaz Marije Krucifikse Kozulić 3, 51000 Rijeka, Croatia,
info@intechopen.com or visit our website at intechopen.com.

We are IntechOpen, the world's leading publisher of Open Access books

Built by scientists, for scientists

5,600+
Open access books available

138,000+
International authors and editors

175M+
Downloads

156
Countries delivered to

Our authors are among the
Top 1%
most cited scientists

12.2%
Contributors from top 500 universities

WEB OF SCIENCE™

Selection of our books indexed in the Book Citation Index (BKCI)
in Web of Science Core Collection™

Interested in publishing with us?
Contact book.department@intechopen.com

Numbers displayed above are based on latest data collected.
For more information visit www.intechopen.com

Meet the editor

Dr. Wei Wu is an associate professor and associate department chair in the Department of Toxicology, Nanjing Medical University, China, where he received his Ph.D. in Toxicology in 2012. He was a guest researcher at the National Institute of Environmental Health Sciences (NIEHS) between 2017 and 2018. Dr. Wu is a member of different national and international societies in the fields of human reproduction and toxicology and has received awards from many national societies for the originality and quality of his projects. Dr. Wu has authored seventy-three peer-reviewed papers in international journals. He has edited four books and collaborated on ten others as well as seventeen patents and in the organization of three international conferences. He is a reviewer for ninety-eight journals.

Contents

Preface

The male reproductive system functions mainly in the production, nourishment, and temporary storage of spermatozoa, which is produced via spermatogenesis. In spermatogenesis, which is a complex, multistep differentiation process, millions of mature spermatozoa are produced by a fertile male daily. This process comprises a variety of unique genetic and epigenetic mechanisms that eventually generate haploid sperm, which provides half of the genetic material and epigenetic information that is needed to create a new life upon fertilization. This book focuses on the male reproductive system, including aspects of anatomy, risk factors related to male infertility, and clinical techniques and management of male reproductive health.

Chapter 1 outlines male reproductive anatomy/embryology and its functions. Testes, ductus deferens, epididymis, accessory glands, and the penis make up the male reproductive system.

Chapter 2 discusses the role of the testes in the production and functions of testicular androgens as well as testicular protein hormones. These hormones maintain the health of the testes and ensure their proper functioning regarding sperm production and delivery.

Chapter 3 describes the structure of the seminiferous tubules of the testes and what occurs inside these tubules, including cell communication and germ cell development from spermatogonia to spermatozoon. This chapter is useful for biologists and physicians working in the field of assisted reproduction to understand the physiology and pathology of spermatogenesis.

Chapter 4 discusses the significance of various positional relationships of male reproductive organs in insects and how this relates to their morphology and function, with a focus on sperm.

Chapter 5 examines chemical agents, physical factors, and biological intruders and their respective effects on reproductive function and potential.

Chapter 6 describes a mouse model for studying the male reproductive system and applies the Johnsen score system to assess testicular histopathology in the seminiferous tubule cross-section. This chapter indicates the negative impact of heat stress on mouse spermatogenesis as well as the human reproductive system.

Chapter 7 discusses sperm selection methods for sperm preparation in assisted reproductive technologies.

Chapter 8 discusses the etiological factors, diagnosis, and treatments of penile necrosis, a complication of post-circumcision.

Chapter 9 overviews the role of epigenetics in normal sperm and epigenetic changes in male infertility. The sperm epigenome is believed to be affected by several biological factors, environmental exposures, and lifestyle. This chapter also discusses environmentally disrupted epigenetic modifications.

Chapter 10 overviews the role of miR-107 and its utility as a possible blood biomarker in prostate cancer, which is the second most commonly occurring cancer in men.

This book would not have been possible without the efforts of numerous contributors. I would like to thank Ms. Marica Novakovic at IntechOpen for her strong support from the inception to the completion of this work. I would also like to acknowledge my coauthors for their efforts.

Wei Wu
Department of Toxicology,
School of Public Health,
Nanjing Medical University,
China

Section 1

Male Reproductive System

Chapter 1

The Concept of Male Reproductive Anatomy

Oyovwi Mega Obukohwo, Nwangwa Eze Kingsley, Rotu Arientare Rume and Emojevwe Victor

Abstract

The human reproductive system is made up of the primary and secondary organs, which helps to enhances reproduction. The male reproductive system is designed to produce male gametes and convey them to the female reproductive tract through the use of supportive fluids and testosterone synthesis. The paired testis (site of testosterone and sperm generation), scrotum (compartment for testis localisation), epididymis, vas deferens, seminal vesicles, prostate gland, bulbourethral gland, ejaculatory duct, urethra, and penis are the parts of the male reproductive system. The auxiliary organs aid in the maturation and transportation of sperm. Semen is made up of sperm and the secretions of the seminal vesicles, prostate, and bulbourethral glands (the ejaculate). Ejaculate is delivered to the female reproduc¬tive tract by the penis and urethra. The anatomy, embryology and functions of the male reproductive system are discussed in this chapter.

Keywords: AMH, TDF, SRY, SF1, DHT etc.

1. Introduction

Reproduction refers to the production of new offspring, also known as breeding in animals. It includes a set of physiological processes (usually) that take place in the female reproductive system with the association of behaviors and anatomical structures that are necessary in order to ensure the birth of the next generation of human, domestic, wild, and laboratory vertebrate organisms. Although these processes take place within the female's system, it is as a means of the fusion of haploid gametes each from male (sperm cell) and female (ovum) termed, fertilization in vertebrates. Testes, ductus deferens, epididymis, accessory glands, and penis make up the male reproductive system [1].

The males' reproductive system functions mainly in the production, nourishment and temporary storage of male gametes (spermatozoa), which is produced via spermatogenesis. It produces androgens and estrogen through steroidogenesis [1] and very importantly, connected to the organ of copulation (penis) which serves to introduce semen containing spermatozoa into the female genital system via mating.

2. Embryology of the reproductive system

The primordial germ cells have shifted from their previous extra embryonic position to the gonadal ridges by the six weeks of development in both sexes, where they are surrounded by the sex cords to form a pair of primitive gonads. The forming gonad, whether chromosomally XX or XY, is potential until this point. The current theory is that the development of an ovary or testis is determined by the synchronized action of a series of genes that contribute to the development of the ovary when there is no Y chromosome or there is no Y testicular development. Unless a gene on the shooting arm of the Y named TDF (testis defined factor) acts as a switch, the ovarian pathway is followed, diverting development into the male pathway.

One of the leading current concerns in medical genetics is the search for the main testis-determined gene. The medullary tissue forms traditional testes with seminiferous tubules and Legdig cells in the presence of the Y chromosome that become capable of androgen secretion under the stimulation of human chorionic gonadotropin (HCG) from the placenta. Spermatogonia, produced by 200 or more successive mitoses from the primordial germ cells, forms the walls of the seminiferous tubules along with the supporting sertus cells. The gonad, by default, produces an ovary if no Y chromosome is present; the cortex develops, the medulla regresses, and oogonia starts to develop within follicles. Oogonia is obtained from primitive germ cells by a sequence of approximately 30 mitoses, less than the number necessary for spermatogenesis.

Oogonia joins meiosis 1 at about the end of the third month, but this process is interrupted at a point called dictyotene, in which the cell persists until ovulation happens several years later. Many of the oogonia degenerate before birth, and during the 30 years or so of sexual maturity of the female, only about 400 mature into ovas. Thickenings in the ridges suggest the developing genital ducts, the mesonephric (formerly called Wolffian) and paramesonephric (formerly called mellerian) ducts, while the primordial germ cells are migrating to the genital ridges. In the male, androgen is released by the Legdig cells of the fetal testes, which stimulates the mesonephric ducts to form the male genital ducts, and Sertoli cells produce a hormone that suppresses paramesonephric duct formation. The mesonephric ducts regress in the female (or in the non-gonadic embryo) and the paramesonephric ducts develop into the female duct system. The outer genitals consist of a genital tubercle, paired labio scrotal swellings and paired urethra folds in the early embryo. Under the influence of androgens, male external genitals develop from

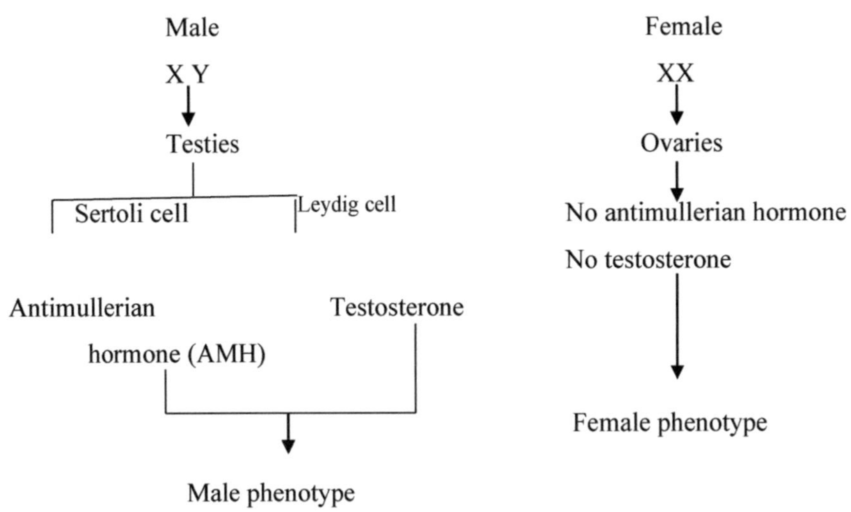

Figure 1.
Sexual differentiation in male and female.

this undifferentiated state or, in the absence of a testis, female external genitals are produced regardless of whether an ovary is present. The male and the female phenotype is as discuss below (**Figure 1**).

3. Male phenotype

- Fetal testicular cells secrete ample testosterone to increase blood concentrations to the same degree as those seen in adult males. Accumulation of testosterone is increased by an additional influence of the gene product TDF gene or SRY (sex determining region of the Y chromosome), which inhibits aromatase production and prevents the conversion of testosterone to estrogens. Testosterone promotes the growth and differentiation of the wolffian ducts that develop into the internal male genital tracts.

- Sertoli cells in the newly differentiated seminiferous tubule secrete a glycoprotein called antimullerian hormone (AMH) under the influence of the SRY gene product and various transcription factors, inducing apoptosis of tubular epithelial cells and atrophy or reabsorption of the mullerian ducts (which would have become the female internal genital tract).

- The primitive structures that give rise to the outside genitalia in both sexes are the urogenital sinus and genital tubarcle. Masculanization of these structures relies on the secretion of testosterone by the fetal testis to form the penis, scrotum and prostate gland. Those structures grow into the female external genitalia unless stimulated by androgen. Differentiation is incomplete when there is insufficient androgen in male embryos or too much androgen in female embryos and the external genitals are unclear. Male external genitalia distinction relies on dilydrotestosterone rather than testosterone.

4. Female phenotype

- Estrogen is secreted by the ovaries in gonadal females but not by testosterone antimullerian hormone.

- Wolffian ducts cannot distinguish without testosterone.

- Mullerian is not suppressed without antimullerian ducts and thus develops into the female internal genital tract.

5. Male reproductive system

The human male reproductive system is a collection of organs that contribute to the reproductive process situated outside the body and around a male's pelvic region. The key direct function of the male reproductive system is to supply the ovum for fertilization by the male gamete or spermatozoa. The male reproductive system is divided into four main compartments (as indicated in **Figure 2**):

1. The testis.

2. Accessory ducts: This includes Epidydimis, Vas deferens, Ejaculatory Duct.

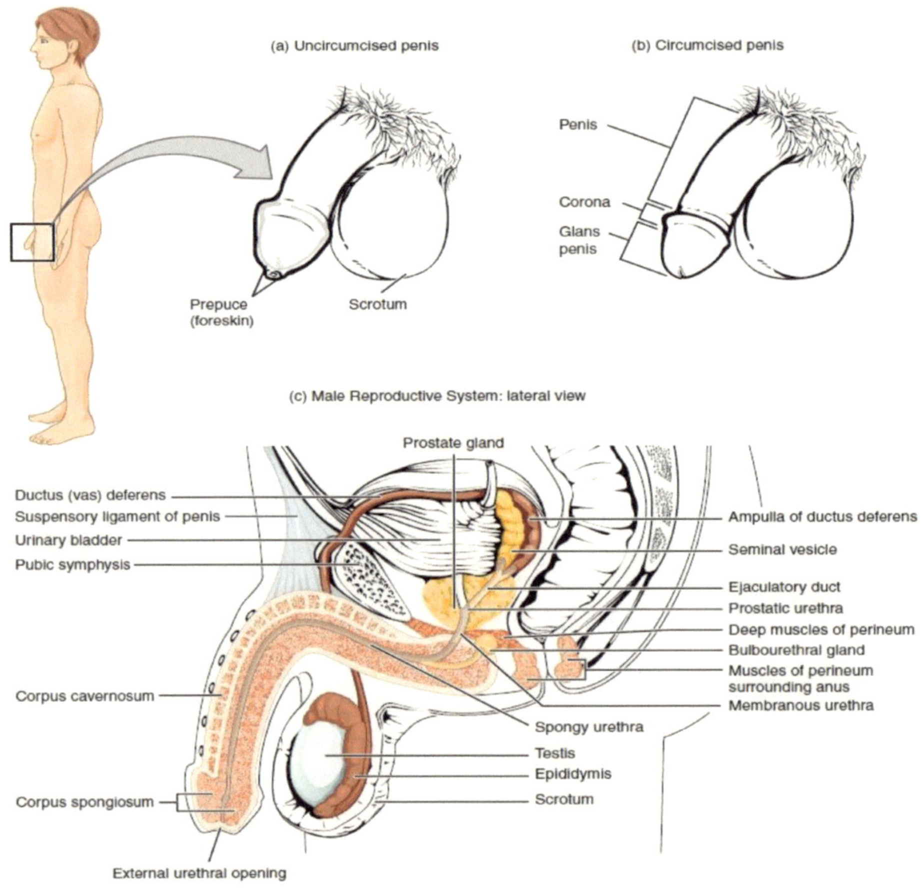

Figure 2.
Typical structure of the male reproductive system.

3. Accessory glands: Accessory glands are internal reproductive organs which supply fluids that nourish the sperm cells and lubricate the duct system. They are the seminal vesicles, the the bulbourethral glands, and the prostate glands (Cowper glands).

4. Supporting structures which include the scrotum and the penis.

In mammals, paired testes, epididymides, ductus deferens, accessory sex glands and penis are part of the male reproductive system. Tests perform two major roles that are very crucial for life perpetuation, spermatogenesis and steroidogenesis [2]. Within the seminiferous tubules of the testis, spermatogenesis or spermatozoa development takes place and steroidogenesis or testosterone synthesis occurs within the interstitial compartment. Spermatogenesis takes place within the stratified epithelium in the seminiferous tubules, while testosterone production takes place inside the Leydig cells that are spread between the seminiferous tubules in a vascular, loose connective tissue in the interstitial compartment [3]. In determining male secondary sexual characteristics, sperm development and fertility, testosterone, developed by the Leydig testis cells, plays an important role [4].

Epididymis is a single, long and extremely convoluted duct that connects the vas deferens (a coiled duct that connects epididymis to the ejaculatory duct) to the testicular efferent ducts. In the transport and storage of testicular spermatozoa, epididymis plays an significant role. Epididymis is categorized in most mammals into three distinct regions on the basis of its gross morphology; caput or head, corpus or body and region of cauda or tail. The area of the corpus is thinner and it joins the larger segments, caput and cauda. There is an additional canal in reptiles between the testes and the epididymis head, which receives the numerous efferent ducts. However, in both birds and mammals, this is missing [5]. A pseudostratified epithelium surrounds the epididymis. The epithelium is divided from the connective tissue wall, which has smooth muscle cells, by a basement membrane. In the epithelium, the main cell types are:

Principal cells: Columnar cells, with much of the epithelium in the basal cells. They also have non-motile stereo cilia, which are long and branching in the head region and shorter in the tail region, extending from the lumen to the basal lamina. [5]. Carnitine, sialic acid, glycoproteins, and ghycerylphosphorylchlorine are also secreted into the lumen as well.

Basal cells: shorter, pyrmid-shaped cells that, before their apical surfaces enter the lumen, touch the basal laminal but taper off [5]. These are known to be undifferentiated primary cell precursors.

Apical cells: These are predominantly located in the head region [5]

Intraepithilial lymphocytes: distributed throughout the tissues.

Clear cell: Predominant in the tail region. The clear cells in the rat epididymis are subdivided into two types and are concerned with the secretion of either glycoproteins or glylipoproteins. The blood epididymal barrier, constituted by the zona occludens of the functional complexes at the apical ends of principal cells [5] also appears to play a vital role in maintaining a physiological millieu in the epididymal canal suitable for sperm maturation.

Spermatozoa formed in the testis are functionally immature and as they migrate through the epididymis they attain functional maturity. Epididymal epithelium absorptive and secretory behavior helps to maintain a particular intraluminal environment that is necessary for sperm maturation [6]. They transfer into the vas deferens, where it is processed before ejaculation, as spermatozoa mature. Sperm flows from the lower portion of the epididymis (which acts as a storage reservoir) during ejaculation. They have not been activated by prostate gland products and are unable to swim, but are transported inside the vas deferens by the peristaltic action of muscle layers and are combined before ejaculation with the diluting fluids of the seminal vessels and other accessory glands. There are some apical variations in the epithelial cells of the epididymis that are sometimes referred to as stereo cillia, as they appear like cilia under the light microscope. However, as electron microscopy has shown that they are more similar to microvilli structurally and functionally, some now refer to them as stereovilli. Stored sperm remain fertile for 40 to 60 days, but they disintegrate and the epididymis resorbs them if they become too mature without being ejaculated. A thin tube approximately 43.2 centimeters long that begins from the epididymis to the pelvic cavity is the vas deferens, also known as the sperm duct. In order to transfer sperm, there are two ducts which connect the left and right epididymis to the ejaculatory ducts. Each tube (in humans is about 30 centimeters long and surrounded by smooth muscle.

The smooth muscle in the walls of the vas deferens contract reflexively during ejaculation, thereby propelling the sperm forward. This is often referred to as peristalsis. The sperm is passed into the urethra from the vas deferens, gathering secretions from the male accessory sex glands, such as the seminal vesicles, the prostate gland, and the bulbourethral glands that make up the majority of the semen. The rate of fluid transfer by the vas deferens is not known in humans. The testes are brought up close to the abdomen just before ejaculation, and fluid is rapidly transferred through the vas deferens into the area of the ejaculatory ducts and then into the prostatic urethra. Intravasal fluid is transported back into the epididymis after ejaculation and even sometimes into the seminal vesicles [7]. Videoradiography during ejaculation after vasography has recorded the retrograde transport of sperm to the seminal vesicles. For some men after vasectomy, the return of sperm to the seminal vesicles after ejaculation can help to explain the prolonged presence of sperm in the ejaculate. The vas deferens can be obstructed or entirely missing, causing male infertility (the latter a possible characteristic of cystic fibrosis). Testicular sperm extraction (TESE), extracting sperm cells straight from the testicles, will resolve it. Seminal vesicles (glandulae vesiculosae) or vesicular glands are paired sac-like or simple tubular glands attached near the base of the bladder to the vas deferens [8]. They are glands of approximately 10 to 15 cm in length that are extremely convoluted [8, 9]. Tubular alveoli with active secretary epithelium are composed of seminal vesicles. The inner surface of the seminal vesicles consists of tubules that form irregular diverticula and are thrown into an intricate system of folds. The main portion of seminal fluid, the fluid that carries spermatozoa, is around 50–80% of the seminal vesicle secretions [10]. A large proportion of the substance that eventually becomes semen is secreted by the seminal vesicles. Dead epithelial cell lipofuscin granules give the scretion its yellowish hue.

Seminal vesicles are highly androgen dependent and contain prostaglandins, proteins, amino acids, citrate, fructose, flavins, enzymes, vitamin C and phosphoryl choline and their secretions are alkaline.

When processed in semen in the laboratory, the high fructose content provides nutrient energy for the spermatozoa. Seminal vesicle secretions enhance sperm capacity, increase sperm stability and help prevent sperm immune response in the female reproductive tract [11]. Alkaline secretion helps to neutralize the vaginal tract 's acidity, thus increasing sperm lifespan [8]. Secretion of the seminal vesicle in semen also tends to improve sperm chromatin stability. In addition, from the fructose found in the seminal secretion, spermatozoa acquire their key energy source.

Prostate is a fibromuscular elastic, donut shaped gland covering the urethra inferior to or at the urinary bladder neck [6]. A thin vascularized fibroelastic tissue layer [6] encapsulates the prostate. It is roughly 2 × 3 × 4 cm in diameter and weighs approximately 20 g. From these endodermal cells, the glandular epithelium of the prostate differentiates and the related mesenchyme differentiates into the prostate 's thick, solid and smooth muscle [12]. The primary function of the prostate is to secrete milky fluid containing proteins and hormones that are part of the seminal fluid produced by seminal vesicles. The prostate fluid is rich in phosphate acids, citric acid, fibrinolysin, antigen specific to the prostate, amylase, callikrein, zinc and calcium, which are essential for spermatozoa to function normally. The secretions of the prostate make up 30 percent of the amount of seminal fluid. The prostate is an androgensensitive organ and relies on the presence or absence of circulating androgens for growth and regression.

Two small glands situated on the sides of the urethra just below the prostate gland are the bulbourethral glands, often referred to as Cowper glands. These glands create a transparent, slippery fluid that directly empties the urethra. They are homologous in female to the Bartholin glands [13]. Compound tubulo-alveolar glands, each about the size of a pea in humans, are the bulbourethral glands [13]. They are made of several lobules with a fibrous covering kept together. Each lobule consists of a number of acini, lined by columnar epithelial cells, opening into a duct that forms a single excretory duct joining the ducts of other lobules. This duct is about 2.5 cm long and opens up at the base of the penis into the urethra. With advancing age, the glands decline steadily in size. Each gland causes a clear, salty, viscous secretion known as pre-ejaculate during sexual arousal. This fluid helps to lubricate the urethra to move through spermatozoa, neutralizing traces of urethra acidicurine [7], and helps to flush out any residual urine or foreign matter. Since there is no sperm in the preejaculate, it is possible for this fluid to absorb sperm, stay in the urethral bulb from previous ejaculations, and conduct it until the next ejaculation. Some amount of prostate specific antigen (PSA) is also produced by the Cowper's gland, and Cowper's tumors can increase PSA to a level that makes prostate cancer suspected [7].

6. Fundamental component of male reproductive anatomy

The male reproductive anatomy is divided into five components which are very fundamental to human reproductive health. These include;

1. **Gonadal development:** At eight weeks of gestation (perios of preganacy), Y chromosomes synthesis of H-Y antigen occurs. In the male, this H-Y antigen causes undifferentiated sex glands to devlop into testes while in female, lack of H-Y antigen causes undifferentiated sex glands to develop into ovaries.

2. **Duct development:** In this case, both sexes start out with two system such as mullerian ducts which develops into fallopian tubes, uterus, inner vagina; Wolffian duct which develops into epididymis, vas differens and seminal vesicles. In the developmental processes, the male fetal leydig cells of the testes secretes sufficient testosterone as those seen in aduct men while the sertoli cells of the testes secretes the antimullerian hormone (AMH). The testosterone (androgen) so secreted is responsible for male sex differentiation during embryogenesis (9th and 13th weeks of pregnancy) and its accumulation is enhanced by an additional effect of the testes determining factor (TDF) gene or *sex determining region of the Y chromosome* (*SRY*) gene product which blocks the expression of aromatase, thus preventing the conversion of testosterone to estrogen. The testosterone thereby stimulate the growth and differentiation of the wolffian ducts, which develop into the male internal genital tracts. However, under the influence of the SRY gene product and specific transcription factors, sertoli cells in newly differentiated seminiferious tubule secretes a glycoprotein called antimullerian hormone (AMH), which causes apoptosis of tubular epithelia cells and atrophy or reabsorption of the mullerian ducts 9 (Which would have become the female internal genital tract). Notwithstanding, the downstream of genes that makes up the *SRY* gene product includes the SOX9 and steriodogenesis factor (SF1). These classified products stimulate the differentiation of sertoli cells and leydigs in the testes and also in the formation of tunica albuginea.

3. **External genital development:** There are two primitive structures of the reproductive anatomy that give rise to the external genitalia in both sexes. This includes the genital tubercle and the Urogenial sinus. However, the masculinisation of these structures to form the penis, scrotum and prostate gland depends on the secretion of testosterone by the fetal testes unless stimulated by androgens; these structures develop into the female external genitalia (clitoris, labia, vagina opening etc). When there is insufficient androgen in male embryos or too much androgen in female embryos, differentiation is incomplete and the external genitalia are ambiguous. Differentiation of the masculine external genitalia depends on the dehydrotesterone rather than testosterone. The mechanism through which this occur is via the conversion of testosterone into dihydrotestosterone (DHT) by an enzyme called 5α-reductase

$$\text{testosterone} \xrightarrow{\text{5?alpha reductase}} \text{dihydrotestosterone} \tag{1}$$

4. **Brain development:** Sex hormone such as testosterone and estradiol exert their influence during development of the fetus. Testosterone secreted into the blood reaches the brain and gets converted into estradiol by an enzyme called Aromatase. The estradiol is what actually help in the masculanization of the human brain. In the female the estradiol secreyed by the ovaries binds to a particular protein called α-fetoprotein and therefore prevent its entering into the brain to protect the female brain from being masculanized by estradiol.

$$\text{testosterone} \xrightarrow{\text{Aromatase}} \text{estradiol} \tag{2}$$

5. **Neural development:** Neural development is one of the earliest systems to begin and the last to be completed after birth. This development generates the most complex structure within the embryo and the long time period of development means in utero insult during pregnancy may have consequences to development of the nervous system. The early central nervous system begins as a simple neural plate that folds to form a neural groove and then neural tube. This early neural is initially open initially at each end forming the neuropores. Failure of these opening to close contributes a major class of neural abnormalities (neural tube defects). Within the neural tube stem cells generate the 2 major classes of cells that make the majority of the nervous system: neurons and glia. Both these classes of cells differentiate into many different types generated with highly specialized functions and shapes.

7. The physio-anatomy of the testes

In adult males, the testis is a strong oval-shaped male gonad, about 4 cm long and 2.5 cm wide in size. Testes are located in the scrotum that regulates its temperature below the normal body temperature to approximately 23°C [1, 14]. There are normally two testis, each weighing about 11–17 g with the right one usually slightly larger and heavier than the left one weighing about 963 + 0 [1]. A testis (singular) is surrounded by a saccular extension, called tunica vaginalis, of the peritoneum inside the scrotum. Underneath the tunica vaginalis, Tunica albuginea is contained and forms the testis' white fibrous capsule [8]. Tunica albuginea is subsequently thickened, assembling the testis mediastinum from which the fibrous septa penetrates the testis and divides into about 200 to 300 wedge-shaped lobules.

There are one to four tightly coiled seminiferous tubules in each testicular lobule where sperm is produced [8]. Testis seminiferous tubules consist of two main types of cells, the germ cells and the supporting cells or Sertoli cells. In the seminiferous epithelium, the Sertoli cells are uniformly distributed along with developing germ cells and they nourish the germ cells during their growth. A basal lamina, which includes peritubular myoid cells, lines the seminiferous tubule. Myoid cells constitute a barrier of partial permeability by preventing large molecules from entering the germinal epithelium. The close and gap junctions that exist between the adjacent Sertoli cells, however, form the main exclusion barrier. The seminiferous epithelium is divided into two distinct compartments by these inter-Sertoli cell junctions, called the blood testis barrier: the basal and the adluminal compartments. Spermatogonia and early spermatocytes live in the basal compartment and are readily available for systemic circulation. The adluminal compartment is sequestered from the systemic circulation, containing meiotic and post-meiotic spermatocytes, and is only exposed to the components transported by Sertoli cells [9]. The undifferentiated spermatogonia that reside in the basal compartment of the seminiferous epithelium undergo a series of mitotic divisions during the process of spermatogenesis to form primary spermatocytes.

The primary spermatocyte is then moved to the adluminal compartment and this requires comprehensive restructuring of the inter-Sertoli closed junctions. The spermatocytes undergo two consecutive meiosis rounds in the adluminal compartment to form mature haploid spermatids. In addition to offering physical support to germ cells, Sertoli cells provide a special atmosphere in the adluminal compartment, which is responsible for transporting sperm from the testis to the epididymis by providing a specialized testis. Development factors and nutrients that are essential for the survival of germ cells are important functions of the testis [15]. Germ cell variables, on the other hand, also play an important role in regulating the behavior of the Sertoli cells. For successful spermatogenesis, the interactions between germ cells and Sertoli cells are important. The interstitial compartment of the testis are made of steroid-secreting Leydig cells, blood and lymphatic vessels, nerves, macrophages, fibroblasts and loose connective tissues. However, the principal cells of this compartment are the Leydig cells of interstitial.

8. Epithelial cells of the testes

8.1 Sertoli cells (also called substantial cells)

Sertoli cells are large, irregularly shaped somatic cells. Sertoli cells are bound by tight junctions to each other at their base. Sertoli cells, as shown by their close contact, are essential to the formation of germ cells. A Sertoli cell can be connected to as many as 6 to 12 spermatids. Sertoli cells assist in the spermiation process, where the final detachment of mature spermatozoa into the seminiferous tubule lumen takes place [16]. Excess cytoplasm resulting from the transition of spermatids to spermatozoa, as well as damaged germ cells, are also targeted and phagocytized by sertoli cells. Moreover, for germ cells, the Sertoli cells also provide structural support and nutrition, secreting fluid. In the seminiferous epithelium of adult rats, the columnar cells stretching from the basal to the luminal compartment are found to occupy a volume of approximately 17–19 percent. Sertoli cell secretes inhibin, which is a gonadal-origin nonsteroidal pituitary receptor [9]. The tight junctions around the circumference of each tubule that lead to the blood-testis barrier were created by a continuous layer of non-germinal Sertoli cells. Via the cytoplasm of Sertoli cells, molecules from the blood join germinal cells. A protein called androgen-binding protein is also secreted into the lumen of the seminiferous tubules by Sertoli cells. The Sertoli cell cytoplasm spreads from the periphery to the tubule lumen and envelops the developing germ cells. It helps to protect the seminiferous tubules from immune attack; on the surface of T lymphocytes, the Sertoli cells generate FAS ligand that binds to the FAS receptor. In this way, by inducing apoptosis of T lymphocytes, it avoids the immune attack of the developing sperm [3]. Sertoli cells refer to the testes' somatic cells, which are important for testes to develop and also for spermatogenesis. Via direct interaction and regulation of the environment inside the seminiferous tubules, these cells (Sertoli) promote the progression of germ cells to spermatozoa. The blood testes barrier (BTB), which is produced near the basement membrane by adjacent Sertoli cells, acts as a "gatekeeper" to prevent harmful substances from reaching germ cells, especially during postmeiotic spermatids. The BTB also divides the seminiferous epithelium into the basal and luminal (apical) compartments to allow the growth of postmeiotic spermatids, namely spermiogenesis, to take place in the apical compartment behind the BTB in a specialized microenvironment. The BTB also contributes to the immune privilege status of the testis, at least in part, so that anti-sperm antibodies against antigens that are transiently expressed during spermatogenesis are not produced [10]. Sertoli cells have become incredibly difficult to remain morphologically stable because during the 14 phases of the epithelial cycle they have a continuously evolving, three dimensional relationship with growing germ cells. There have been many Sertoli cell functions identified, most of which are directly related to the production and movement of germ cells. These include 1) the provision of structural support; 2) the production of an impermeable and immunological barrier; 3) involvement in the movement and spermation of germ cells; 4) nutrition of germ cells through their secretory products [10].

8.2 Leydig cells (or interstitial cells of the leydig)

Leydig cells are polygonal in form and are the main type of cell inside the interstitial tissue where they are mostly located adjacent to the seminiferous tubules and blood vessels. Other cell types, such as fibroblasts, macrophages and a limited number of mast cells, are also present in the interstitial space, in addition to Leydig cells. The primary source of testosterone in the systemic circulation of males is the Leydig cells. The Leydig cell cytoplasm contains a lot of mitochondria, a granular endoplasmic reticulum, lipid droplets and occasionally some protein crystals [10]. Leydig cells do not have follicle stimulating hormone (FSH) receptors. Therefore their growth is influenced indirectly rather than directly by the FSH. FSH activates the Sertoli cell development growth stimulators, which in turn stimulated the growth of the Leydig cells that were growing. In addition, the proliferation of developing Leydig cells can also be stimulated by the androgens. However, proliferation and activity of these cells are reduced by the Estrogen receptors that are present in the Leydig cells. Leydig cells have LH receptors, and inducing androgen secretion through a cAMP-dependent mechanism is the main effect of the luteinizing hormone (LH). Testosterone is the primary product of Leydig cells, but dehydroepiandrosterone (DHEA) and androstenedione, two other androgens of less biological activity, are also a product of Leydig cells [17]. However, now that human testes live in the scrotum, they have adapted to this cooler climate and are unable to generate sperm at the 37°C core body temperature. There are three mechanisms in the scrotum to control test temperature:

Cremaster muscle: The cremaster muscle consists of strips that enmesh the spermatic cord of the internal abdominal obligue muscle. The cremaster contracts and pulls the testicles closer to the body when it is cold to keep them warm. The cremaster relaxes when it is warm and the testicles are suspended further from the body.

Darto muscle: A subcutaneous layer of smooth muscle is the darto muscle (tunica dartos). When it is cold, it, too, contracts, and the scrotum becomes taut and wrinkled. The scrotum 's teaching helps to keep the testes snugly against the warm body and decreases the scrotum's surface area, thus decreasing heat loss.

The pampiniform plexus is an extensive network of veins in the spermatic cord from the testes that surround the testicular artery. These converge as they pass through the inguinal canal to form the testicular vein, which emerges into the pelvic cavity from the canal. Warm arterial blood will heat the testicles and prevent spermatogenesis without the pampiniform plexus. However, by serving as a countercurrent heat exchanger, the pampiniform plexus avoids this. Such a process in the spermatic cord eliminates heat from the descending arterial blood, so this blood is 1.5c to 2.5c cooler than the core body temperature by the time it enters the testicles.

Most fish and amphibians do not have seminiferous tubules. The sperm is instead formed in the spherical form known as sperm ampullae. These are seasonal structures, which during the breeding season release their material and are then reabsorbed by the body. Fresh sperm ampullae begin to develop and ripen before the next breeding season. In higher vertebrates, with the same variety of cell types, the ampullae are otherwise virtually similar to the seminiferous tubules.

9. Testicular temperature regulation of testes

The testes perform best at temperature slightly lower than the core body temperature. At lower and higher temperatures, spermatogenesis is less effective [11]. This is possibly why the testicles are found outside of the body. To hold the tests at the optimum temperature, there are various mechanisms [11].

10. Testicular development

The germ cells migrate from the yolk sac to the genital ridge during the 3rd week of development after fertilization. In male embryos, testes develop from the genital ridges from the 4th to the 8th week, and primordial germ cells migrate from the wall of yolk sacs to the gonads. The Leydig cells of the developing testis are starting to evolve under the influence of human chorionic gonadotropin. Testosterone is secreted. The labioscrotal swellings merge at around week 9 to form the scrotum. In order to form the epididymis, vas deferens and seminal vesicles, testosterone also induces mesonephric (Wolfian) duct production [11]. The gubernaculum shortens and pushes the testes, the deferent duct, and its vessels downward between the 7th and the 12th week. The testes remain in the area of the inguinal canal between the 3rd and 7th months so that they may enter into it. Under the control of the androgen hormone, they enter the scrotum at roughly the time of birth. The vaginal process appears as an outpouching of the parietal peritoneum at about 13 weeks of development. The testis stays for 10 to 12 weeks at the beginning of the vaginal process, the internal inguinal ring. This patent herniation mechanism is at least partially dependent on the musculature of the abdominal wall to produce an elevated intra-abdominal input. The patent processus vaginalis does not advance through the inguinal canal if the abdominal muscles are unable to raise intra abdominal pressure, and the testis may not descend into the scrotum. Each testis is formed from three sources: First, in the 7th week of intrauterine life, the production of testes becomes apparent. The medulla of the undifferentiated genital ridge, and the cortex of which regresses, is the base of each testis. The proliferation of coelomic mesothelium covering the medial surface of the mesonephric ridge forms the genital ridge. From the proliferation of the endoderm of the dorsal wall of the hind intestine, primitive sex cells or gonocytes are produced and appear in the genital ridge through active dorsal and sephalic migration between the primitive dorsal mesentery layers of the gut. From the surface of the genital ridge, multiple solid cellular testis cords emerge and project into its interior. Within the testis cords, primitive sex cells are inserted. A cellular plexus and the rete cord, which is located near the blind ends of the mesonephric tubules, are connected by the inner ends of the testis cords to form. Invading the genital ridge, the mesenchymal cells of the mesonephrc ridge spread under the surface, later disconnecting the peripheral ends of the testis cords from the surface. Tunica albuginea forms this portion of the invaded cells. Some of the mesenchymal cells between the testis cords project inwards and persist as a septa testis, and the interstitial cells are formed from the mesenchymal cells that are detached. The testis cords and rete cords are canalized during the 7th month of intrauterine life and form the seminiferous tubules and the rete testis, respectively. Secondly, efferent testis ductules are created to form the proximal 12–15 of the persistent menosephric tubules that form secondary ties with the rete testis. Epididymis and vas deferens are formed from the mesonephric duct in the third channel. The epididymis precedes the testis into the processus vaginalis at 26 to 36 weeks of growth. These structures descend into the scrotum and are fused with the scrotum 's posterior layers, providing an anchor that prevents the movement of the testis. The vaginal process closes at 37 to 40 weeks (full term), preventing all contact between the peritoneum and the inguinal canal or scrotum. A proximal remnant (or more than one remnant) may persist as a small appendage, the appendix epididymis, as the mesonephric duct evolves into the epididymis. Most frequently, this tissue is connected to the caput (most proximal and cephalad portion) epididymis. Such an appendix can sometimes twist and become inflamed. The paramesonephric structures (Müllerian) simultaneously regress under the influence of the Müllerian inhibiting substance (MIS) secreted from the developing testis by the Sertoli cells.

11. Blood neurovascular supply of the testes

The survival of the cells in target organs depends on the delivery of nutrient-rich, oxygenated blood and the removal of metabolic waste. In addition, neural signaling is required for most organs to perform specific duties. In terms of the testes, each of the spherical reproductive organs is supplied by a rather basic bilateral neurovascular network. The extensive vascular supply of the testes serves a variety of functions in addition to supplying oxygen, nutrients, and eliminating waste from the area. This is due to the organs' temperature-sensitive functionality, as well as their dual roles as endocrine glands and reproductive organs.

11.1 Arterial supply

The testicular arteries are a pair of arterial structures on either side of the abdominal aorta that branch straight from it. They arise at the level of the base of the L1-L2 vertebra from the anterolateral surface of the massive artery caudal to the renal vessels. The right testicular artery travels inferolaterally, medial to the right testicular vein and the proximal section of the right ureter, after crossing the inferior vena cava anteriorly. The artery crosses the ureter anteriorly and continues its inferior path on the body of the psoas major. The left testicular artery runs medial to the testicular vein on the left side.

In comparison to the right testicular artery, it has a more vertical proximal path. It also passes anteriorly through the left ureter. The common and external iliac vessels are served by both the left and right testicular arteries. Only when they enter the inguinal canal via the deep inguinal ring do they cross the external iliac vessels (at which point the external iliac vessels become the femoral vessels). They run lateral to the vas (ductus) deferens and its artery within the canal. The testicular artery gives a branch to the epididymis after it enters the scrotum before bifurcating into lateral and medial branches. These two branches further split to perforate the organ's material directly. There are also three noteworthy vascular anastomotic connections formed with the testicular artery. Each of the cremasteric arteries originates on the anteromedial side of its corresponding inferior epigastric artery (branch of the external iliac artery) and forms an anastomosis with the testicular artery as it passes through the spermatic cord (in the inguinal canal). The inferior vesical artery, which is supplied by the anterior segment of the internal iliac artery, gives birth to the ductus deferens artery. It also connects to the testicular artery via an anastomosis.

11.2 Venous drainage

Around the testicular artery, a venous plexus is formed by a dense network of connected veins. The pampiniform plexus is a network that travels cranially with cooler, deoxygenated, nutrient-poor blood. The plexus' branches continue to consolidate as it leaves the scrotum and enters the spermatic cord, eventually becoming four branches. Two branches join at the deep inguinal ring, on either side of the testicular artery. As a result, each testicular artery has two valvular testicular veins that run alongside it to their drainage locations. The two veins then merge to produce a single testicular vein that flows laterally alongside the testicular artery across the psoas muscles anterior surface. The neurovascular supply to the testes is definitely not a light topic, but interactive anatomy can definitely make it easier to study. Each testicular vein crosses its corresponding ureter on the front surface of the psoas muscles about the level of the L3 vertebra. The left testicular vein then travels almost vertically to pierce the left renal vein, passing between the testicular artery on the medial side and the ureter on the lateral side. The right testicular vein, on the other hand, goes practically vertically on the left side, then obliquely on the right side (also with the ureter lateral and the testicular artery medial) before draining straight into the inferior vena cava.

11.3 Innervation

The sympathetic nerve fibers that innervate the testes come from the T10 spinal segment. The lesser splanchnic nerves carry them to the celiac ganglion, where they synapse. The testicular artery is then followed along its route to its place of innervation by the post-ganglionic fibers. Sensory root fibers follow a similar path, passing information to the T10 segment's dorsal root ganglion cells. The testes' tunica vaginalis receives sensory innervation from the genital branch (L2) of the genitofemoral nerve (L1, L2) of the lumbar plexus.

11.4 Lymphatic drainage

The testes are the only structures in the male external genitalia that do not leak into the inguinal lymph nodes. Its lymphatics follow the path of the testicular veins until they reach the para-aortic lymph nodes at the L2 vertebral level.

12. Male reproductive functions

The male reproductive organs are specialized for the following functions:

- Spermatogenic function; for sperm production,
- maintainance and transport of sperm (the male reproductive cells and protective fluid semen)
- Sperm Discharge function; for discharging sperm inside the female reproductive tract.
- Hormonal function; for producing and secreting male sex hormones like testosterone.

12.1 Spermatogenic functions

12.1.1 Semen

Sperm cells and secretions of the seminal vesicles, prostate, Cowper's gland and, perhaps, urothral glands are included in the fluid that is ejaculated in time of organism. It has a fixed gravity (1.028), a bright, opalescent fluid and a PH of 7.35–7.50 of it. For each ejaculation, the approximate volume of semen is 2.5 to 3.5 ml after several days of consistency [18]. The seminal vesicles contain the bulk of this secretion or fluid (about 60 percent), and the prostate gland contributes the remainder (about 40 percent). Components of seminal vesicle secretion include fructose, phosphorylcholine, ergothioneine, ascorbic acid, flavins and prostaglandins, while spermine, citric acid, cholesterol phospholipids, fibrinolysis, fibrinogenase, zinc, and acid phosphate are components of prostate secretion. Semen is also known to contain buffers (phosphate and bicarbonate) and hyaluronidase. The volume of the semen containing sperm decreases rapidly with repeated ejaculation. The sperm in human males ranges between 60 and 150million per millimeter in the ejaculated semen (which accounts for about 20% of the semen volume), even though it takes just only one sperm to fertilize the ovum. Human sperm moves through the female genital tract at a rate of about 3 nm/min and reaches the uterine tubes 30–60 minutes after copulation (sexual intercourse).

A sperm concentration below about 10 millimeter is termed oligospemia, and is associated with decreased fertility. Various factors, including heat from a sauna or hot tub, various prescription medications, lead and assenic poisoning and illicit drugs such as marijuana, cocaine and anabolic steroids, may cause oligospermia. In addition to low sperm counts, some men and women have antibodies against sperm antigens as a cause of infertility (this is very common in men with vasectomy). These antibodies do not tend to influence well being; however they decrease fertility. Secretion from the epididgmins, seminal vesicles, prostate gland and bulbourethral glands along with sperm composition makes up just 1 percent of the semen or seminal fluidsperm, the rest is made up of accessory gland fluids. Semen is over 90% water but contains many substance, most notably energy rich fructose, the known vitamins which include Vitamins C and inositol and the trace elements which include Calcium, Zinc, Magnesinm, copper and sulfur. Semen also contains the highest concentration of prostaglandin in the body. The consistency of semen varies from thick and viscous to almost watery fluid. Primordial germ cells are the first cells destined to become semen. They are produced in the sac of the yolk, a membrane connected with the embryo that is developing. They move into the embryo itself in the fifth to sixth week of development and colonize the seminiferous tubule, beyond the blood-test barrier (BTB). By mitosis, spermatogonia multiplies, producing two types of type A daughter cells and type B spermatogonia. Type A cells remain beyond the barrier of blood tests and begin to multiply from puberty until death. Therefore, men never exhaust their supply of gametes and typically remain fertile in old age. Spermatogonia type B migrates closer to the lumen of the tubule and differentiates into slightly large cells known as primary spermatocytes. These cells must pass through the membrane of the blood testicles and travel into the tubule lumen.

The tight junction between two sustentacular cells is usually dismantled ahead of the primary spermatocyte, while a new tight junction is forms on the other side. The primary spermatocyte undergoes mitosis 1, which gives rise to two equal- size, haploid secondary spermatocytes. Each of these undergoes meiosis II, dividing into two spermatid or a total of four for each spermatogomia. Each stage is a little bit closer to the tubule than the previousr stages. All stages on the luminal side of the blood testies barrier are bound to the sustentacular cells by the tight junctions and gap junction and are closely enveloped in tendrils of the sustentacular cells. Throughout this meiotic division, the daughter cell, remain connected to each other by means of narrow cytoplasmic bridges and do not completely separated. Hence, the rest of spermatogenesis is called spermiogenesis. It does not involve further cell division, but a gradual transformation of each spermatid (immature sperm) into a matured spermatozoon.

12.1.2 Spermatogenesis

The cellular divisions and developmental changes that occur within the seminiferous tubules of the testes are termed spermatogenesis, and it consists of two major parts (**Figure 3**). In part 1, spermatocytogenesis occurs in which it starts with spermatogonia which involve mitotic division of stem cells to form spermatocytes that take place in the early stage, followed by meiosis where the number of chromosomes is reduced to form spermatids. In part 2, spermiogenesis occurs in which the spermatids are transformed in regards to metamorphic changes to sperm [19]. Spermatogenesis is a highly organized but complex process and it normally continuous throughout life [20]. The above description categories spermatogenesis into

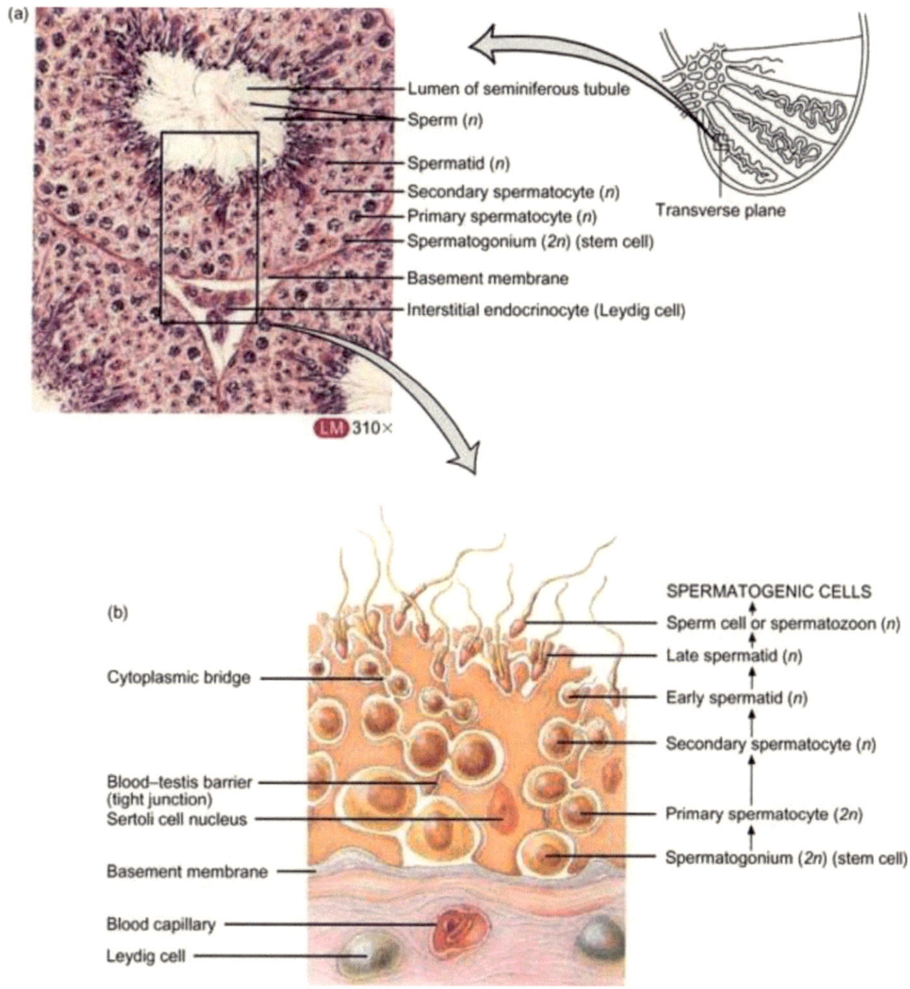

Figure 3.
Showing the Microscopic anatomy of the seminiferous tubules.

major three divisions; spermatocytogenesis, meiosis and spermiogenesis respectively. The process begins from spermatogonial stem cells that are found on the basement membrane of the seminiferous tubules, which usually proliferate for self-renewal and reproduced to a progeny of the differentiating spermatogenic cells such as (1) primary spermatocytes, (2) secondary spermatocytes, (3) spermatids and (4) spermatozoa [21]. The spermatogonia are duplicated mitotic division, one of the duplicate member called primary spermatocyte undergoes meiotic division in order to form secondary spermatocytes. When the spermatogonia (which are a diploid primary spermatocyte) complete the first meiosis, two daughter haploid cells will be produced, a result which is known as secondary spermatocytes. By the end of the second (2nd) meiotic cell division, each of the two (2) secondary spermatocytes formed two (2) haploid spermatids [3]. In the beginning, the spermatids will still pose the normal characteristics of epithelioid cells, however, they differentiate and elongate into matured spermatozoa. A matured spermatozoon comprises of a tail and a head which contains a condensed nuclear material, a thin cytoplasm and a surrounding membranous layer [22]. The major features of spermiogenesis includes the formation of the acrosome derived from the Golgi apparatus, condensation, elongation of the nucleus, formation of the flagellum and extensive shedding of the cytoplasm of the spermiated, spermatozoa consists of a head, middle piece and tail (**Figure 4**) [23].

12.1.3 Structure of mature spermatozoa and its membrane

The sperm consists of a head, a centerpiece, and a tail. The head comprises nuclei surrounded by an acrosome of tightly packed chromatin. The acrosome

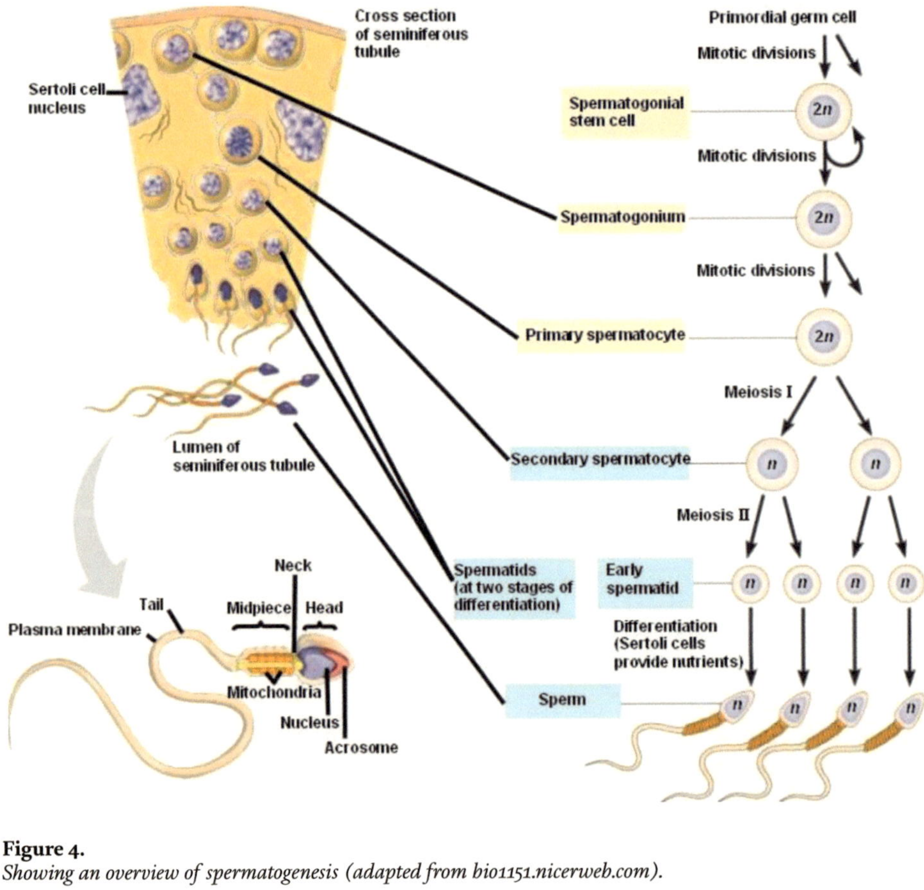

Figure 4.
Showing an overview of spermatogenesis (adapted from bio1151.nicerweb.com).

includes enzymes that are used for oocyte penetration. A special arrangement of mitochondria spiraling around the middle part of the sperm is used for the production of ATP for the passage of the sperm through the female reproductive tract. Spermatozoa are driven by the tail or flagellum of the spermatozoa. Axoneme is the microtubule and related protein bundle that forms the center of the flagellum of eukaryotic sperm and is responsible for movement. Sperm cells and secretions of the seminal vesicles, prostate, Cowper's gland and, perhaps, urothral glands are included in the fluid that is ejaculated in time of organism.

It has a fixed gravity (1.028), a bright, opalescent fluid and a PH of 7.35–7.50 of it. For each ejaculation, the approximate volume of semen is 2.5 to 3.5 ml after several days of consistency [18]. The seminal vesicles contain the bulk of this secretion or fluid (about 60 percent), and the prostate gland contributes the remainder (about 40 percent). Components of seminal vesicle secretion include fructose, phosphorylcholine, ergothioneine, ascorbic acid, flavins and prostaglandins, while spermine, citric acid, cholesterol phospholipids, fibrinolysis, fibrinogenase, zinc, and acid phosphate are components of prostate secretion.

Semen is also known to contain buffers (phosphate and bicarbonate) and hyaluronidase. The volume of the semen containing sperm decreases rapidly with repeated ejaculation. The sperm in human males ranges between 60 and 150million per millimeter in the ejaculated semen (which accounts for about 20% of the semen volume), even though it takes just only one sperm to fertilize the ovum. Human sperm moves through the female genital tract at a rate of about 3 nm/min and reaches the uterine tubes 30–60 minutes after copulation (sexual intercourse)

A sperm concentration below about 10 millimeter is termed oligospemia, and is associated with decreased fertility. Various factors, including heat from a sauna or hot tub, various prescription medications, lead and assenic poisoning and illicit drugs such as marijuana, cocaine and anabolic steroids, may cause oligospermia. In addition to low sperm counts, some men and women have antibodies against sperm antigens as a cause of infertility (this is very common in men with vasectomy). These antibodies do not tend to influence wellbeing, however they decrease fertility. Secretion from the epididgmins, seminal vesicles, prostate gland and bulbourethral glands along with sperm composition makes up just 1 percent of the semen or seminal fluidsperm, the rest is made up of accessory gland fluids. Semen is over 90% water but contains many substance, most notably energy rich fructose, the known vitamins which include Vitamins C and inositol and the trace elements which include Calcium, Zinc, Magnesinm, copper and sulfur. Semen also contains the highest concentration of prostaglandin in the body. The consistency of semen varies from thick and viscous to almost watery fluid. Primordial germ cells are the first cells destined to become semen. They are produced in the sac of the yolk, a membrane connected with the embryo that is developing. They move into the embryo itself in the fifth to sixth week of development and colonize the seminiferous tubule, beyond the blood-test barrier (BTB). By mitosis, spermatogonia multiplies, producing two types of type A daughter cells and type B spermatogonia. Type A cells remain beyond the barrier of blood tests and begin to multiply from puberty until death. Therefore, men never exhaust their supply of gametes and typically remain fertile in old age. Spermatogonia type B migrates closer to the lumen of the tubule and differentiates into slightly large cells known as primary spermatocytes. These cells must pass through the membrane of the blood testicles and travel into the tubule lumen.

The tight junction between two sustentacular cells is usually dismantled ahead of the primary spermatocyte, while a new tight junction is forms on the other side. The primary spermatocyte undergoes mitosis 1, which gives rise to two equal- size, haploid secondary spermatocytes. Each of these undergoes meiosis II, dividing into two spermatid or a total of four for each spermatogomia. Each stage is a little bit closer to the tubule than the previousr stages. All stages on the luminal side of the blood testies barrier are bound to the sustentacular cells by the tight junctions and gap junction and are closely enveloped in tendrils of the sustentacular cells. Throughout this meiotic division, the daughter cell, remain connected to each other by means of narrow cytoplasmic bridges and do not completely separated. Hence, the rest of spermatogenesis is called spermiogenesis. It does not involve further cell division, but a gradual transformation of each spermatid (immature sperm) into a matured spermatozoon (**Figure 5**).

During sperm passage through the epididymis, spermatozoa collected or derived from the testis do not show progressive motility or capacitate, but developed these abilities [24]. Dynamic morphological and metabolic changes leading to the development of active sperm capable of fertilizing the ovum are referred to as sperm maturation. These processes are called maturation. The completion of nuclear condensation and changes in the distribution and expression of molecules on the surface of the sperm are all part of sperm maturational changes. Phospholipid hydroperoxide glutathione peroxidase (GPx4) may be used as an alternative reductant to glutathione in the sperm nucleus by the thiol groups in nuclear proteins. ROS lipid peroxide generation could provide GPx4 with a substrate to drive the oxidation of these proteins and promote nuclear condensation, while providing protection against oxidative DNA damage at the same time [25]. By enhancing cyclic adenosine monophosphate (cAMP) synthesis and protein phosphorylation at the time of ejaculation, reactive oxygen species could also be involved in motility initiation [26]. For successful fertilization, the membrane structure of spermatozoa plays a pivotal role, as both the acrosome reaction and sperm-oocyte fusion are membrane-associated events; in fact, the spermatozoa membrane lipids are essential for spermatozoa fluidity and flexibility. These lipids, however, along with membrane proteins, are also the key substrates for peroxidation that can cause serious sperm functional disorders [27]. High oxidant concentrations have been shown to provoke sperm pathology such as ATP depletion, leading to inadequate axonemal phosphorylation, lipid peroxidation and loss of motility and viability. The adverse influence of reactive oxygen species (ROS) is due to the sperm plasma membrane's peroxidative damage. In addition, in a high proportion of infertility patients, oxidative stress-mediated damage to the sperm plasma membrane can account for defective sperm function observed. In spermatozoa maturation, capacitation and the initiation of the gamete interaction process, ionic environment and ionic fluxes through the membrane are extremely significant. In the mammalian sperm plasma membrane, various kinds of ion channels are found, indicating a number of different functions in sperm physiology and gamete interaction.

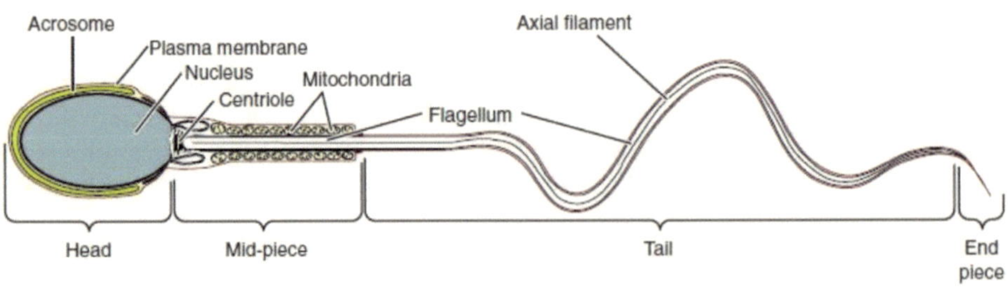

Figure 5.
Structure of a mature spermatozoon (adapted from [24]).

The plasma membrane integral enzymes in most animal cells are Na^+/K^+-ATPase (E.C. 3.6.1.9) and Ca^{2+}-ATPase (E.C. 3.6.1.3) and are important components involved in ionic homeostasis. Changes in the surrounding of the sperm membrane and thus in fluidity change the activities of these enzymes, requiring the existence of phospholipids closely linked to their structure. The Na + pump is a heteromeric protein consisting of several isozymes and is not only responsible for maintaining cell osmotic equilibrium, volume and pH, but also for maintaining the capacity of the cell resting membrane and supplying chemical energy across the cell membrane for the secondary Na + −coupled transport of other ions, solutes and water. The Ca^{2+} pump, on the other hand, is responsible for the homeostasis of calcium that is central to normal cell function. In particular, a distinctive Na^+/K^+-ATPase isoform expression profile has been found in the mammalian testis with regard to the Na^+ pump. Sanchez *et al.* stated that the human Na^+/K^+-ATPase 4 isoform has different functional properties and plays a primary role in the motility of sperm. Sulphydryl (SH) containing enzymes are considered to be both ATPases and their thiol groups may be the target for both nitric oxide (NO) and its derivatives such as peroxynitrite (ONOO-). In fact, it has been clearly shown that NO and NOderived reactive nitrogen species modulate the activity of different enzymes and can thus damage cells, causing sperm dysfunction by increasing lipid peroxidation, complete depletion of the sulphydryl group and formation of nitrotyrosine or by inactivating proteins, damaging nucleic acids, which in turn leads to to alteration or disturbances in membrane structure and function. The development of several disease states in humans is the accumulation of this oxidative damage and, in particular, the progressive oxidation of sperm thiols to disulphides is involved in sperm chromatin condensation and stabilization of the tail structure needed for the subsequent initiation of motility. In particular, it has been revealed that peroxynitrite inhibition of Na^+/K^+-ATPase activity is followed by a reduction in the number of protein thiol groups and a shift in the enzyme's substrate dependency curve [28, 29]. This means that the blockade of Na^+/K^+-ATPase ATPase SH-groups is responsible for its inhibition [29]. The pattern of this inhibition is consistent either with the oxidation of thiol groups directly involved in the binding of ATP but in a way that cannot be resolved by raising the concentration of the substrate ('noncompetitive') or with the oxidation of SH groups located outside the enzyme's active site but essential for the enzyme's activity.

12.1.4 Capacitation and acrosome reaction

Capacitation is a morphological transition that spermatozoa are subjected to by hyperactivation and acrosome reaction sequence to gain the capacity to fuse with an ovum [30]. Sperm motility hyperactivation is characterized by a high amplitude, asymmetrical sperm tail beating pattern and enables the sperm to enter the ovum zone pellucida. It is accompanied by the acrosome reaction where acrosin and other enzymes are released by the head of the mature spermatozoa to digest the cumulus cells and break through the zona pellucida [24]. Research has shown that O2-plays an extremely important regulatory function in promoting both hyperactivated motion and acrosome reaction induction [24, 31]. Increased membrane fluidity, increased tyrosine phosphorylation, increased pH levels, increased intracellular cAMP, and calcium influx are characterized by capacitation. Substances present in semen, progesterone, peroxiredoxin-4 and other substances secreted by the oocyte cumulus complex [29] can regulate capacitation, but can also occur spontaneously under sufficient in vitro conditions. Moreover, through the redox regulation of tyrosine phosphorylation, ROS produced by mammalian spermatozoa can play a physiologically important role in driving the complex process of capacitation.

The mechanism that support this redox-effect on protein tyrosine phosphorylation include a number of additional signal transduction stalls, such as sarcoma and extracellular kinase regulated signal mediation pathways, stimulation of inhibition of tyrosine phosphatase activity by cAMP generation. Capacitation is therefore carried out by increasing membrane fluidity, cholesterol efflux, ion fluxes leading to sperm membrane potential alteration, increased protein phosphorylation of tyrosine, hyperactivation induction, and acrosome reaction. Reactive oxygen species function alongside other factors including bicarbonate, membrane cholesterol loss, and increased intracellular Ca^{2+} resulting in activation of the cyclase of adenyl (AC), leading to cAMP production and activation of protein kinase A (PKA) and the phosphorylation of tyrosine proteins. Lewis and Aitken proposed that adenyl cyclase is activated by superoxides, while Rivlin *et al.* [32] proposed that cyclase is activated by hydrogen peroxides that may substitute for bicarbonate. Increased cAMP activates PKA, which activates tyrosine kinases and, by unknown mechanisms, inhibits tyrosine phosphatassase (TP). The participation of PKA with the PKA inhibitor (H89) was confirmed. Hydrogen peroxide stimulates TK directly and inhibits TP. The main driving force of capacitation and conduct to hyperactivation, zone binding and acrosome reactions is the increase of the tyrosine phosphorylation induced by such changes [33]. The increase of fertilization by 50 percent by induction of mild LPO using a mixture of ferrous ion and ascorbic acid was seen in vitro studies performed on mouse sperm. The study shows that a strongly hydrogen peroxide-induced OS activates sperm activity and improves fertilization rate. Superoxide anion induces capacitation in incubation conditions by the effects of an oxidase. Further stimulation of the development of ROS; superoxide anion, hydrogen peroxide induce the release from plasma of these cells of unesterified fatty acid.

12.1.5 Sperm abnormalities

Sperm anomalies, which are usually based on sperm concentration, motility, and morphology, include: oligospermia (sperm concentration lesser than 20 million/ml). This is supported by Iammarrone *et al.* [34], which showed low conception rate in human sperm counts with a concentration lesser than 20 million/ml. A complete lack of spermatozoa in an ejaculate is termed azoospermia and such is found to accounts for 10–15 percent of male infertility cases. Partial obstruction of sperm duct also influenced sperm concentration [35]. Asthenospermia (poor sperm motility), is a condition in which spermatozoa are too slow in movement, not able to strive in a straight line along the cervical mucus within the female reproductive tract and or fertilize the egg. When 60% or more sperm actively move in a straight line, the percentage motility is said to be normal, and quality is at least average. In cases where percentage motility is less than 40%, the sperm the condition is as less qualitative. Genetic or otherwise sperm defects may be responsible for sluggishly sperm movement and this may render them incompetent of fertilizing the egg. Poor sperm motility associated with DNA fragmentation can increase the risk of genetic diseases transmitted to offspring. Sperm motility is rated in two ways: percentage of the total motility (general motility), or the individual forward progressive sperm movement (progressive motility) [36]. The latter is a grade dependent on the pattern of the majority of motile sperm. It ranges from null indicating (no movement) to four (suggesting excellent forward progression). Notably, a sperm sample needs to have at least 50% progressive motility. Teratospermia or morphologic abnormalities are usually categorized based location of the deformity of a spermatozoon, whether it is on the head, neck (midpiece), or tail.

Primary and secondary anomalies are the most important classification scheme types: primary abnormalities are structural defects in the location affecting head, midpiece and tail. While the sperm was still inside the seminiferous epithelium of the testis, a more primary serious defect is thought to originate while secondary defects are considered less extreme and thought to occur during the passage through the epididymis or by mishandling after ejaculation (sperm). The heterogeneous state of teratozoospermia includes changes in the form of various components of sperm. There is a strong connection between morphological defects and the potential for sperm fertilization, since mature spermatozoa structures have the best organization to serve specific functions. Teratozoospermia can therefore be considered to be a mixture of morphological defects with associated sperm function impairments [36].

12.1.6 Factors influencing spermatogenesis

The spermatogenesis process is highly sensitive to environmental fluctuations, especially hormones and temperatures. In order to sustain the process, which is accomplished by bidding testosterone with androgen binding protein present in the seminiferous tubules, testosterone is needed at large local concentrations. Testosterone is produced by interstitial cells that reside adjacent to the serminiferous tubule, also referred to as ledig cells. In humans and certain other animals, the seminiferous epithelium is susceptible to elevated temperatures and can be adversely affected by temperatures as high as average body temperature; therefore, the testes are found in a skin sack called the scrotum outside the body. At 20°C (Man) -80°C (mouse) below body temperature, the optimum temperature is preserved. This is accomplished by controlling blood flow and by placing the cremasteric muscle and dartos smooth muscle towards or away from the body heat. A nutritional deficiencies (such as vitamins B, E, and A), anabolic steriod, metals (Cadmium and lead), X-ray exposure, dioxin, alcohol, drug toxicant and diseases of pathogens may also adversely affect the rate of spermatogenesis [29, 37].

12.2 Testicular steroidogenesis

For both spermatogenesis and the development of secondary sex characteristics, steroidogenesis, which involved the production of testosterone (T) and dihydrotestosterone (DHT) from cholesterol by a series of P450 enzymes in the Leydig testis cells, is essential. The differentiation of the Wolffian ducts into the epididymides, vasa deferentia, seminal vesicles, and the development of the levatorani-muscle and bulbocavernosus gland (the LABC complex) is the responsibility of T in utero (produced locally by the interstitial Leydig cells regulated by LH). DHT (produced locally in the testis by T conversion using the 5-alpha-reductase enzyme) is responsible for differentiating the genital tubercle from the external genitalia and the urogenital sinus into the glands of the prostate and Cowper and for regression of nipple anlagen in the male fetuses. According to the receptors to which they attach, steroid hormones can be classified into five distinct groups: mineralocorticoids, glucocorticoids, androgens, estrogen and progestagen. Cholesterol, the basic precursor for biosynthesis of all steroid hormones, is integrated by receptor-mediated endocytosis into the Leydig cell from low-density lipoproteins or is synthesized de novo from acetate within the cell. In cytoplasmic lipid droplets, cholesterol is contained in an ester form and the number of droplets in Leydig cells is regarded to be inversely proportional to the rate of androgen synthesis [38]. LH-induced cholesterol ester hydrolase activation hydrolyzes cholesterol ester during steroidogenesis, which is transported into the mitochondria of Leydig cells. The StAR protein is used to transport cholesterol from the outside to the inner mitochondrial membrane. The exact mechanism by which cholesterol is transported by StAR protein to the mitochondria, however, remains uncertain. StAR protein is regulated acutely, and protein expression is critically dependent on stimulation of trophic hormones (e.g. LH and ACTH). This makes it sensitive to toxicants from the environment: several xenobiotics [e.g. 4-tert-octylphenyl and pesticides Lindane (1,2,3,4,5,6-hexachloro-cyclohexane) and glyphosate Roundup (2-(phosphonomethylamino) acetate)] have been reported to interfere with StAR protein expression inhibitor Steroidogenesisby [39, 40]. The condition lipoid congenital adrenal hyperplasia (lipoid CAH) is believed to be caused by mutations in the StAR gene. Lipoid CAH is an autosomal recessive lethal condition in which cholesterol and cholesterol esters accumulate and a sufficient amount of steroids can not be synthesized by the newly born child. In humans, StAR knockout mice display a phenotype that is very similar to lipoid CAH, providing a clear model for studying the mechanism of the important contribution of StAR protein to steroidogenesis and endocrine production. In the inner mitochondrial membrane, the cytochrome P450scc side chain cleavage enzyme, which belongs to the monooxygenase family, transforms cholesterol to pregnenolone. Three successive monooxygenations are involved in this step: 22-hydroxylation, 20-hydroxylation and C20-C22 bond cleavage. Pregnenolone then diffuses across the mitochondrial membrane and is translocated to the endoplasmic reticulum, where it undergoes a series of testosterone-forming biochemical reactions. Pregnenolone undergoes C17 hydroxylation in the Delta 5 pathway to form 17 alpha-hydroxypregnenolone, which is then split between C17 and C20 bonds to form DHEA. The cytochrome P450 17 alpha-hydroxylase/ C17, 20 lyase catalyzes these reactions [41]. DHEA could be transformed by the action of 3β-HSD to androstenedione and then 17β-HSD to testosterone. The equilibrium between these androgens depends on the present activity and type of 17b-HSD. Types 3 and 5 of 17b-HSD catalyze the conversion of androstenedione to testosterone and are expressed in Leydig testis cells, while the opposite reaction occurs in type 2 (found among others in prostate and placenta) [42].

In steroidogenic and non-steroidogenic tissue such as asthetes, prostate, skin and brain, the enzyme 3β-HSD is commonly expressed. Four 3β-HSD isozymes exhibiting differential and tissue-specific expression were characterized in the rat. The spermatic vein transports testosterone into circulation. Testosterone synthesis is governed by LH in Leydig cells. Testosterone biosynthesis is also regulated by other factors, such as FSH, insulin-like growth factor-1 and cytokines [43]. By paracrine regulation of testicular functions, FSH also regulates spermatogenesis (**Figure 6**).

12.3 Sperm discharge function

The discharge of semen into the reproductive tract of female has to do with the following steps;

a. **Libido:** This is the biological need (sexual drive) for sexual activity and is often expressed as conduct that seeks sex. Its strength is variable over a given time between people and within an person. In stable older but not younger men, higher serum testosterone tends to be associated with greater sexual action.

b. **Erection:** The enlargement and firm state of penis is the erection of the penis. The dynamic interaction of psychological, neuronal, vascular and endocrine factors are some of the factors it depends on. When two tubular structures running the length of the penis, the corpora cavernosa, are engorged with venous blood, a penile erection occurs. This can result from any of the different physiological stimuli. The corpus spongiosum is a single tubular structure situated just below the cavernosa corpora, which comprises the urethra from which, during urination and ejaculation, urine and semen move through. This may often be slightly engorged with blood, but less so than the penile erection of the

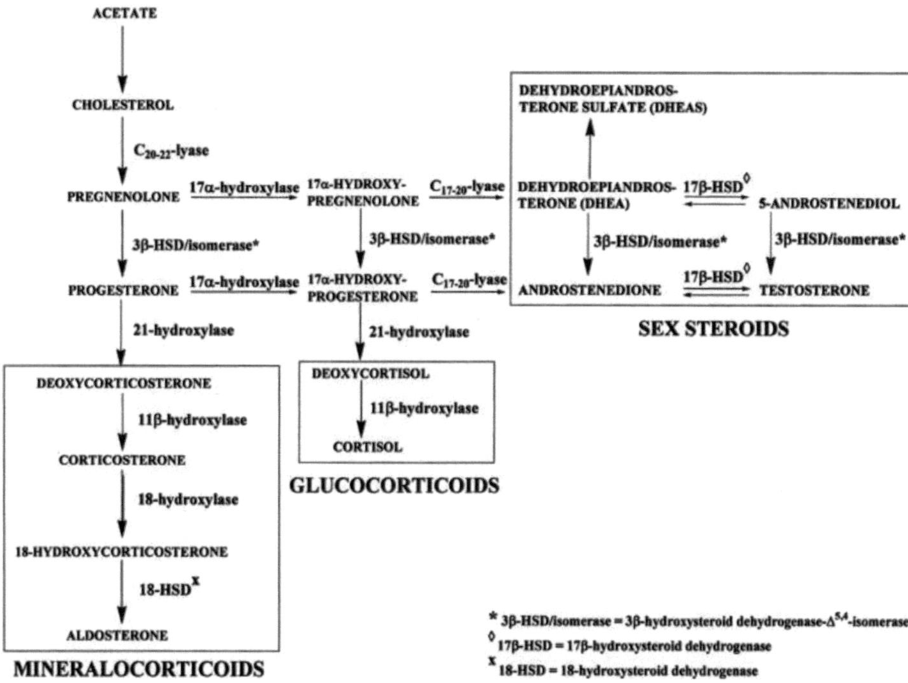

Figure 6.
Showing the major pathways in steroid biosynthesis.

corpora canvernosa normally results from sexual arousal and/or excitement, but can also occur due to triggers such as a full urinary bladder or spontaneously over the course of a day or at night, often during romantic or wet dreams. Swelling and enlargement of the penis results from an erection. Although it is not necessary for all sexual activities, erection makes sexual intercourse and other sexual activities (sexual functions).

c. **Ejaculation:** Ejaculation and erection must take place for sperm to deliver into the female genital tract (without technical assistance). Two occurrences or events should really be considered during ejaculation, the first being semen deposition into the posterior urethra, called seminal emission, and the second being semen expulsion from the urethra. The sympathetic nervous system controls emission and ejaculation of sperm. Emission includes the intense contraction of the vas, ampulla of the vas, seminal vesicles and prostate covering muscle and myoid complexes. Ejaculation, along with contraction of the periurethral muscles, mainly the bulbocavernosus muscle, requires closure of the bladder neck to avoid or prevent retrograde semen flow.

d. **Orgasm:** Organsm, or orgasm, is an intense, pleasurable feeling that typically happens at the height of sexual arousal, accompanied by a decrease in sexual tension. Not all sexual arousal results in orgasm, because in order to have an orgasm, people need various circumstances and different forms and quantities of stimulation. Orgasm is made up of a rhythmic contraction series. In the pelvic organs and genital area. Throughout the body, breathing rate, heart rate, and blood pressure increase dramatically. The general contraction of the muscles can lead to facial contortions and muscle contractions in the extremities, back, and buttocks. Organism occurs in two phases in men. First, at the base of the urethral, the vas deferens, seminal vesicles, and prostate contract, sending seminal fluid to the bulb, and the man feels a sense of inevitability of ejaculation, a sensation that ejaculation is just about to happen or happen and can not be prevented. Second, a mechanism called ejaculation is closely related to the urethra bulb and penis rhythmically contracting, expelling the sperm, but some men experience organism separately from ejaculation.

13. Endocrine and neuroendocrine factors regulating testicular functions

The brain (on stimulation), the master endocrine gland, and local factors generated by the testes finely regulate the spermatogenic and steroidogenic functions of the testes. The proliferation of primitive germ cells and the development of the testes are carefully regulated by testosterone (secreted by the legdig cell on activation by placenta released human chorionic gonadotropin (HCG)) during intrauterine life [44]. The hormonal control of testes ceases after birth and the testes remain quiet until the beginning of puberty [45]. The testicular function setting is triggered at puberty by certain cells in the hypothalamus that activate GnRH secreting cells (**Figure 7**). These cells are referred to as kisspeptin secreting cells found in the periventricular nucleus (PVN), preoptic nucleus (PN) and arcuate nucleus (ARC) and in the anteroventral perivetricular nucleus (AVPV) [44].

Steroids, leptin, and other systemic factors are believed to have effect on the testicular functions by binding on receptor located on these kisspeptin secreting cells [44]. Kisspeptin stimulates GnRH cells to release gonadotropin releasing hormone via the median eminence (**Figure 8**). The cells that secrete GnRH are under the regulation of kisspeptin because they express the GPR54 receptor on their cell membrane that bind to kisspeptin released by kisspeptin secreting cells. This hormone is carried to the anterior pituitary through the hypothalmo-hypophyseal portal system. There the gonadotropes are stimulated by GnRH to release follicle stimulating hormone (FSH) and luteinizing hormone (LH) which are the tropics for leydig and sertoli cells respectively. This is the neuroendocrine axis of testicular regulation. The gonadotropes in the adenohypophysis, on stimulation, releases LH, FSH and growth hormone that regulate the functions of the testes. This is the endocrine axis of

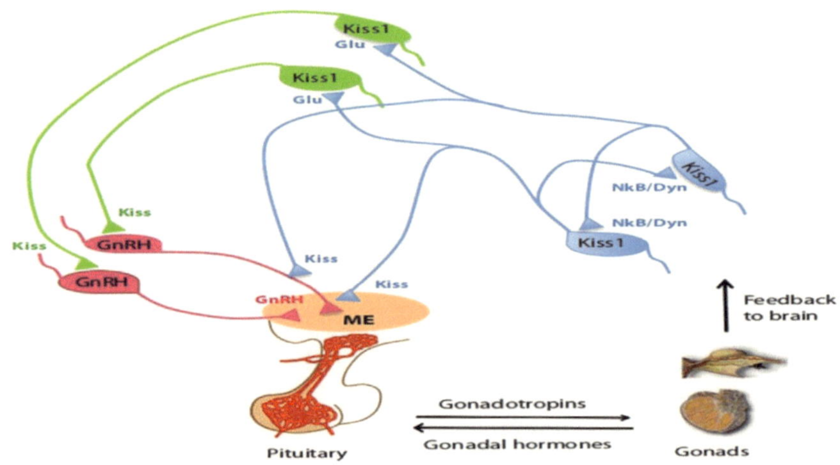

Figure 7.
Showing kisspeptin cells connections with GnRH cells (source: www.wikipedia.org).

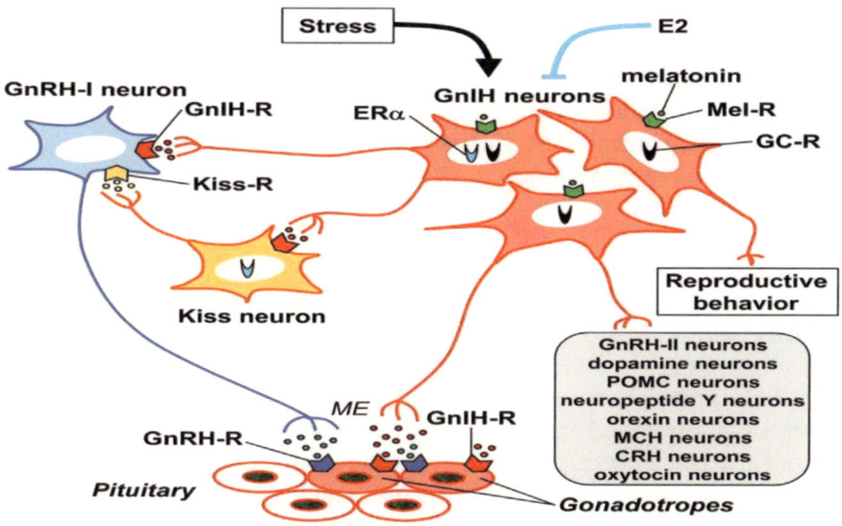

Figure 8.
Regulation of kisspectin-GnRH axis (Wikipedia.org.) It is through this kisspeptin-GnRH axis that factor such as steroids stress, leptin light and dark etc. influence the functions of the testicles [46].

testicular regulation. Spermatozoa development is based on pituitary gonadotropins, LH and FSH, which are released in response to hypothalamic GnRH pulsatile release.

The testes as an endocrine gland secrete steroid and other local factors that regulate its function through autocrine and paracrine mechanism. The steroids and inhibin synthesized by leydig and sertoli cells respectively regulate the neuro-endocrine and endocrine factors via negative feedback mechanism. Abnormalities in these levels of testicular regulation results to male reproductive dysfunctions [29, 37]. This is hypogonadotropin-hypogonadism due to misdirection in GnRH cells migration from the olfactory cells during development [44].

GnRH act via interacting with a particular receptor found on the cell membrane of gonadotropes. These receptors are G-protein - coupled receptors that interact with the hormone to form a hormone receptor complex. This results in the interaction of phosphoinositide with Gp protein hydrolysis and the release of diacylglycerol and inositol triphosphate, resulting in the mobilization of calcium from intracellular stores and the inflow of extracellular calcium into the cell. The release of gonadotropin from gonadotropes into the general circulation results from this calcium influx [44].

14. Clinical implications of gonadotropins and steroidogenic hormones

LH binds to the receptors located on the Leydig cells in the testis and induces testosterone synthesis, which in turn could adversely affect the release of hypothalamic and pituitary hormones. FSH targets the receptors on the Sertoli cells and induces androgen-binding protein production, which helps to transport testosterone via the Sertoli cells' tight junction complexes. Sertoli cells are also activated by FSH to secrete inhibin and activin, both of which have a negative effect on hypothalamus and pituitary hormone release. The primary endocrine hormone involved in testicular function control is FSH. FSH has a central role to play in regulating the Sertoli cell populations, which in turn modulates the number of germ cells proceeding through the mitotic and meiotic spermatogenesis phases. In mitotic and meiotic spermatogonia, FSH handles or regulate DNA synthesis and also prevents apoptosis induction in round spermatids [47]. It has been shown that FSH stimulates the release of various products from Sertoli cells. Sertoli cell products have been reported to play a role in the regulation of the functions of Leydig cells. For successful spermatogenesis and steroidogenesis, the ability of LH to function on the LH receptors present on Leydig cells is essential. LH controls the growth of Leydig cells, the number of Leydig cells, the biosynthesis of testosterone and its secretion. The removal of testosterone has been shown to induce spermatid detachment from Sertoli cells, resulting in full spermatogenesis stoppage. To initiate, sustain, and restore spermatogenesis, testosterone works synergistically with FSH. In particular, testosterone contributes to the blood-testis barrier development, the maintenance of interactions between Sertoli and germ cells, and the release of mature sperm from Sertoli cells). The blood-testis barrier formation is weakened in the absence of testosterone and germ cells are released from the Sertoli cells prematurely.

Estrogens, localized in Leydig and Sertoli testis cells, efferent ductules and epididymis also play a significant role in spermatogenesis control. Evidence suggests that estrogen is secreted into the seminiferous tubular fluid by germ cells, which may be essential for the efferent ductules and epididymis functions. It is stated that estrogen has a stimulatory and inhibitory effect on the proliferation and differentiation of germ cells. It has been shown that administration of aromatase inhibitors to male monkeys induces decreased spermatogenesis and sperm concentrations, suggesting estrogen 's crucial role in sustaining spermatogenesis. In proliferating Sertoli cells, high estrogen levels are present and their levels decrease as Sertoli cells avoid differentiation and start maturation. Estrogens control the expression of the molecule of cell adhesion, neural cadherins, involved in the maintenance of cell adhesion of germ cells-Sertoli.

Environmental estrogens is known to have a deleterious effects on male fertil¬ity and it has been shown that neonatal exposure to exogenous estrogens induces irreversible alteration of gene expression in the reproductive tract. Testicular steroidogenesis in adulthood has been shown to impair neonatal sensitivity to diethylstilbestrol, a synthetic estrogen. Administration of 17β-estradiol to adult rats has been shown to induce a decrease in basal and stimulated testosterone production of 33–48 percent. Adult male rats showed a substantial decrease in circulating FSH

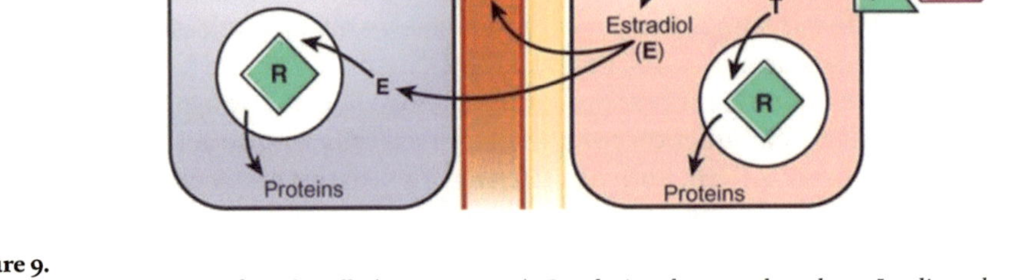

Figure 9.
Showing the main product of Leydig cells (e.g testosterone), Regulation, hormonal products, Leydig and Sertoli cells interaction. AB (androgen binding protein); ATP (adenosine triphosphate); cAMP (cyclic adenosine monophosphate); E (estradiol); FSH (follicle-stimulating hormone); LH (luteinizing hormone); T (testosterone) [17].

and LH concentrations when treated with estradiol, which subsequently contributed to reductions in serum and testicular testosterone levels. Several other factors, apart from hormones, have also been shown to affect testicular functions (**Figure 9**).

15. Other factors necessary for male reproductive functions

- **Growth Hormone:** This hormone is usually synthesized by the anterior pituitary (AP) and has effect on virtually all tissues and organs of the body. It is essential for the general metabolic processes in the testes. It is also necessary for proliferation of spermatogenesis, and plays a key role in the proliferation, growth and maturation of spermatids [45].

- **Local Factors:** Some of the most important local factors produced in the testes by the leydig cells are testosterone and insulin-like factor while that of sertolin cells of thae testes are inhibin, growth factor. Stem cell factors, immunological factors, opioids, oxytocin, vasopressin, peritubular cell modifying factors, rennin, angiotensin, GnRH, CRH, ACTH, GHRH, calmudilin, pasminogen activator, metalloproteases, dinophin, PACAP etc. are some of the other local factors that are believed to have effects on the functions of the testes [44]. Apart from these factors, glucose has also been shown to be important for proper functioning of testis.

- **Insulin signaling and glucose transport:** Apart from these factors, glucose has also been shown to be important for proper functioning of testis. Glucose is very critical for high-energy, challenging testicular spermatogenesis and steroidogenesis to be successfully accomplished. It has been shown that cytochalasin B, a glucose transport inhibitor, competitively binds to proteins that are involved in Leydig cells' facilitated glucose absorption, and inhibits testosterone synthesis stimulated by LH. In the presence of glucose, high testosterone production has been observed, suggesting the need of this compound for testosterone production in addition to LH, and it has also been shown that there is no testosterone production in the absence of glucose. The family of facilitative glucose transporter (GLUT) proteins carries out glucose transfer through the plasma membranes. There are 13 GLUT protein families that have been identified to date. Glucose transporter-1 to −3 expressions in different types of rat testicular cells has been demonstrated. Mature spermatozoa also express glucose transporters because they require glucose for basic cell activity as well as for specific functions such as motility and fertilizing capacity [48]. One of the recently cloned members of the GLUT family, GLUT-8 is known to be the leading transporter of glucose in the testis. In the heart, skeletal muscles, brain, spleen, prostate and intestine, GLUT-8 is expressed, but its expression was found to be highest in the testis relative to all other tissues [49, 50], thus indicating the involvement of GLUT 8 in glucose transport for steroidogenesis of Leydig cells [51–53]. In addition, in testicular cell types, GLUT-2 has also been shown to be abundantly expressed. The high expression in the testis of insulin signaling molecules and glucose transporters suggests the high energy expenditure of contractile testicular cells and the dependency on glucose as an energy source. The insulin receptor family also plays an important role in the development of gonads in the testis. The differentiation of the testis is caused by the sex-determining region Y (SRY) expression present in somatic progenitor cells intended to become Sertoli cells. Sertoli, Leydig, interstitial and myoid cells have been shown to express IRS-1 and IRS-2, suggesting the reliance of these cell types on insulin. Sperm motility, progressive motility and acrosome reaction of human spermatozoa have been reported to increase by insulin and leptin, thereby improving their fertilizing ability. In addition, it has been shown that human spermatozoa releases pulsatile insulin, which is autocrine-regulated, and it has been hypothesized that insulin derived from sperm can play a role in sperm capacitation [54]. Therefore, insulin plays an important role in the proper functioning of the testis and in preserving the capacity of spermatozoa to fertilize. Insulin signaling and glucose transport in the body are known to be affected by many factors. Of the different variables, as one of the main regulators of glucose homeostasis in the body, reactive oxygen species (ROS) are involved. While low ROS levels are important for the signaling of insulin, increased ROS could have a negative impact on homeostasis of glucose.

Author details

Oyovwi Mega Obukohwo[1*], Nwangwa Eze Kingsley[2], Rotu Arientare Rume[3] and Emojevwe Victor[4]

1 Department of Hunan Physiology, Achievers University, Owo, Ondo State, Nigeria

2 Faculty of Basic Medical Science, Department of Physiology, Delta State University, Abraka, Nigeria

3 Department of Physiology, University of Ibadan, Ibadan, Oyo State, Nigeria

4 Department of Physiology, University of Medical Sciences, Ondo, Ondo State, Nigeria

*Address all correspondence to: megalect@gmail.com

IntechOpen

© 2021 The Author(s). Licensee IntechOpen. This chapter is distributed under the terms of the Creative Commons Attribution License (http://creativecommons.org/licenses/by/3.0), which permits unrestricted use, distribution, and reproduction in any medium, provided the original work is properly cited.

References

[1] Stevens, A., & Lowe, J. (2005): Chapter 16: Male reproductive system. *Hum histol (3rd ed.).* (Pp: 327-343).

[2] Carreau, S., Bouraima-Lelong, H., & Delalande, C. (2012): Estrogen, a female hormone involved in spermatogenesis. Adv. Med. Sc., *57(1), 31-36.*

[3] Ridge, I.L.; Dubuque, I.A., & Madison, W.I. (2004): Human physiology/Stuart Ira Fox. 8th ed. McGraw-Hill companies, Americas. Pp: 644-654.

[4] Hess, R. A., Renato de Franca, L. (2008): Spermatogenesis and cycle of the seminiferous epithelium. Adv Exp Med Biol *636, 1-15.*

[5] Kierszenbaum, A. (2002): Sperm axoneme: A tale of tubulin posttranslation diversity. Mol Reprod and Development. *62(1):1-3.*

[6] Cooper, T.G. (2011): The epididymis, cytoplasmic droplets and male fertility. Asian J Androl. *13:130-138.*

[7] Kim, Howard H., Goldstein and Marc. (2010): Chapter 53: Anatomy of the epididymis, vas deferens, and seminal vesicle". In Graham, Sam D, Keane, Thomas E, Glenn, James F. Glenn's urological surgery (7th ed).356.

[8] Akinsola, A. R., Oluwaseun, H., Adewale, A., Olusegun, S., & Adesina, M. (2012): Effect of the methanolic extract of trichosanthescucumerina seed (snakegourd/tomatoe) on hormone influenced seminal vesicle weight in adult Wistar rats. Webmedcentral Anatomy, *3(6), 1-8.*

[9] Mescher, A. L. (2010): Chapter 21: The male reproductive system. Junqueira's basic histology text and atlas (12th ed.). USA: McGraw-Hill Companies.

[10] Kierszenbaum, A. L., & Tres, L. (2015): Histology and cell biology: an introduction to pathology. 4th ed. Elsevier Health Sciences. Pp: 653.

[11] Gonzales, G. F. (2001): Function of seminal vesicles and their role on male fertility. Asian J. Androl. *3(4), 251-258.*

[12] Moore and Persaud (2008): Before we are Born, Essentials of Embryology and Birth Defects, 7th edition. Saunders Elsevier.

[13] McEntee, M. (2012): Reproductive Pathology of Domestic Mammals. Elsevier Science. *333.*

[14] Saladin, K. S. (2008): Chapter 26: Reproductive system. Human anatomy (2nd ed.). (pp. 736-769). New York. McGraw-Hill Companies

[15] Petersen, C. and Soder O. (2006): The Sertoli cell--a hormonal target and 'super' nurse for germ cells that determines testicular size. Horm Res. *66(4): 153-161*

[16] Rhoades, R.A., & Tanner, G.A. (2004): Medical physiology. PART X Reproductive Physiology: the male reproductive system. 2nd edition. Lippincott Williams and Wilkins. Pp: 660

[17] Rhoades, R. A., & Bell, D. R. (Eds.). (2012): Medical Physiology: Principles for clinical medicine. 4th ed. Lippincott Williams & Wilkins. Pp: 681.

[18] Ganong, W.F. (2010): Review of medical physiology, 23rd edition. The McGraw-Hill companies, Inc.

[19] Gribbins, K. M., Happ, C. S., & Sever, D. M. (2005): Ultrastructure of the reproductive system of the Black Swamp Snake (Seminatrix pygaea). V. The temporal germ cell development strategy of the testis. *Acta Zoologica, 86(4), 223-230.*

[20] Ohmura, M., Ogawa, T., Ono, M., Dezawa, M., Hosaka, M., Kubota, Y., & Sawada, H. (2003): Increment of murine spermatogonial cell number by gonadotropin-releasing hormone analogue is independent of stem cell factor c-kit signal. Biol Reprod, *68(6), 2304-2313*

[21] Nakamoto, T., Shiratsuchi, A., Oda, H., Inoue, K., Matsumura, T., Ichikawa, M., & Hirai, H. (2004): Impaired spermatogenesis and male fertility defects in CIZ/Nmp4-disrupted mice. Genes to Cells, *9(6), 575-589.*

[22] Guyton, C. A., & Hall, E. J. (2006): Text Book of Medical Phyiology (11thed.). Philadelphia, Pennsylvania: Suanders.

[23] Vanputte, C., Regan, J., Russo, A.F., Tate, P., Stephens, T.D., Seeley, R.R. (2013

[24] Gadella, B. M., Rathi, R., Brouwers, J. F. H. M., Stout, T. A. E., Colenbrander, B. (2001): Capacitation and the acrosome reaction in equine sperm. Anim Reprod Sci, *68:249-265*

[25] Pfeifer, H.M., Conrad, D., Roethlein, A., Kyriakopoulos, M., Brielmeier, G.W., Bornkamm, D. B. (2001): Identification of a specific sperm nuclei selenoenzyme necessary for protamine thiol cross-linking during sperm maturation. FASEB, *15:1236-1238.*

[26] Aitken, R.J., Ryan, A.L., Baker, M.A., McLaughlin, E.A. (2004): Redox activity associated with the maturation and capacitation of mammalian spermatozoa. Free Radic Biol Med, *36:994-1010.*

[27] Agarwal, A., Gupta, S., Sikka, S. (2006): The role of free radicals and antioxidants in reproduction. Curr Opin Obstet Gyn *18, 325-332.*

[28] Koçak-Toker N, Aktan G, Aykaç-Toker G 2002 The role of Na,KATPase in human sperm motility. Int J Androl *25, 180-185.*

[29] Oyovwi M.O, Tesi P.E, Rotu R.A, Nwangwa N.K, Emojevwe V, Benneth Ben-Azu, Ozegbe Q.E.B. Ameliorative Effects Of Lutein Against Cyclosporine-Induced Testicular Oxidative Stress, Apoptotic And Inflammatory Flux In Male Wistar Rats. Clinical and Experimental Obstetrics & Gynecology,2021, 48(3):733-777 DOI: 10.31083/j.ceog.2021.03.0511

[30] Gaboriau, D., Howes, E. A., Clark, J., Jones, R. (2007): Binding of sperm proacrosin/ßacrosin to zona pellucida glycoproteins is sulfate and stereo dependent.Synthesis of a novel fertilization inhibitor. *Dev.* Biol, *306: 646-657.*

[31] Ferramosca, A., Zara, V. (2014) Bioenergetics of Mammalian Sperm Capacitation. *Rev article biomed res int, 1-8.*

[32] Rivlin, J., Mendel, J., Rubinstein, S., Ektovitz, N., Breitbart, H. (2004): Role of hydrogen peroxide in sperm capacitation and acrosome reaction. Biol Reprod, *70:518-522.*

[33] Breitbart, H. (2003): Signaling pathways in sperm capacitation and acrosome reaction. Cell Molecular Biology, 49:321-327.

[34] Iammarrone, E., Balet, R., Lower, A. M., Gillott, C., & Grudzinskas, J. G. (2003): Male infertility. Best Pract Res Clin Obstet Gynaecol, *17(2), 211-229.*

[35] García-González, F. (2004): Infertile matings and sperm competition: the effect of “nonsperm representation” on intraspecific variation in sperm precedence patterns. The Am Nat, *164(4), 457-472.*

[36] Kishore, K. P. and Raju, A. B. (2011): A Review on male fertility. Hygeia J. Drugs and Medicines, *3(1), 20-28.*

[37] Kunle-Alabi OT, Akindele OO, **Mega O. Oyovwi**, Duro-Ladiipo AM and Raji Y (2014). *Cocos nucifera L.* water improves reproductive indices in Wistar rats. Afr. Jour. Med. Med. Sci. 43: 305-313 (M.Sc Research Thesis)

[38] Chang, L., Xu, J., Zhao, J. (2006): Taurine antagonized oxidative stress injury induced by homocysteine in rat vascular smooth muscle cells. Acta Pharmacol Sin *2004; 25:341-346.*

[39] Walsh, L.P., McCormick, C., Martin, C., Stocco, D.M. (2000). Roundup inhibits steroidogenesis by disrupting steroidogenic acute regulatory (StAR) protein expression. Environ Health Perspect *108 (8):769-776.*

[40] Walsh, LP. and Stocco, D.M. (2000). Effects of lindane on steroidogenesis and steroidogenic acute regulatory protein expression. Biol Reprod *63(4):1024-1033.*

[41] Dharia, S., Slane, A., Jian, M., Conner, M., Conley, A.J., Parker, C.R., Jr. (2004). Colocalization of P450c17 and cytochrome b5 in androgen-synthesizing tissues of the human. *Biol Reprod71(1):83-88.*

[42] Mindnich, R., Moller, G., Adamski, J. (2004): The role of 17 beta-hydroxysteroid dehydrogenases. *Mol Cell Endocrinol218(1-2):7-20.*

[43] Sofikitis, N., Giotitsas, N.T., sounapi, P., Baltogiannis, D., Giannakis, D., Pardalidis, N. (2008): Hormonal regulation of spermatogenesis and spermiogenesis. J Steroid Biochem Mol Biol *109323-*109*330.*

[44] Weinbauer, G.F., Luetjens, C.M., Simoni, M., Nieschlag, E. (2010): Physiology of Testicular Function. In: Nieschlag E, Behre HM, Nieschalag S, editors. Andrology - Male reproductive health and dysfunction. *Berlin Heidelberg: Springer-Verlag. pp. 11-59.*

[45] Sembulingam, K. & Sembulingam, P. (2006): Essential of Medical Physiology. New Delhi, India: Jaypee

[46] Clarke, I, J.(2011). Control of gonadotropin secretion: One step back. *Frontiers in Neuroendocrinol.*

[47] Vigier, M., Weiss, M., Perrard, M.H., Godet, M., Durand, P. (2004): The effects of FSH andof testosterone on the completion of meiosis and the very early steps ofspermiogenesis of the rat: an in vitro study. J Mol Endocrinol. *33:729-742.*

[48] Bucci, D., Rodriguez-Gil, J. E., Vallorani, C., Spinaci, M., Galeati, G., & Tamanini, C. (2010): GLUTs and Mammalian Sperm Metabolism. J. Androl, *32(4), 348-355.*

[49] Ibberson, M., Riederer, B.M Uldry, M., Guhl, B., Roth, J., & Thorens, B. (2002): Immunolocalization of GLUTX1 in the testis and to specific brain areas and vasopressin-containing neurons. Endocrinol.*143276-284*

[50] Ibberson, M., Uldry M., & Thorens B. (2000): GLUTX1 a novel mammalian glucose transporter expressed in the central nervous system and insulin-sensitive tissues. *J. Biol Chem, 2754607-4612.*

[51] Chen, Y., Nagpal, M.L., Lin, T. (2003): Expression and regulation of glucose transporter 8 in rat Leydig cells. J. Endocrinol. *17963-72.*

[52] Gomez, O., Ballester, B., Romero, A., Arnal, E., Almansa, I., Miranda, M., Mesonero, J.E.,, Terrado, J. (2009): Expression and regulation of insulin and the glucose transporter GLUT8 in the testes of diabetic rats. Horm Metab Res. *41: 343 – 349*

[53] Schurmann, A., Axer, H., Scheepers, A., Doege, H., Joost, H.G. (2002): The glucosetransport facilitator GLUT8 is predominantly associated

with theacrosomal region of mature spermatozoa. Cell Tissue Res. *2002;307(2):237-242.*

[54] Aquila, S.; Gentile, M.; Middea, E. (2005): Autocrine regulation of insulin secretion in human ejaculated. Spermatozoa. *Endocrinol, v.146, n.2, p.552-557.*

Chapter 2

Endocrine Functions of the Testes

Emojevwe Victor, Igiehon Osarugue, Oyovwi Mega Obukohwo, Nwangwa Eze Kingsley and Naiho Alexander Obidike

Abstract

The testes, also known as the male gonads are found in the scrotal sacs. In addition to their spermatogenic functions, they also secrete steroids and protein hormones. The steroid hormones are the androgens, testosterone and dihydrotestosterone as well as estrogen, while the protein hormones are inhibins, activins, and anti-Mullerian hormone (AMH). This chapter therefore discusses the role of the testis in the production and functions of the testicular androgens as well as testicular protein hormones.

Keywords: testicular functions, anti-Mullerian hormone (AMH), testosterone, inhibin, activin

1. Introduction

The testes, also known as the male gonads, are male reproductive organs located in the scrotal sacs and basically responsible for spermatogenesis [1]. However, various studies have shown that in addition to the spermatogenic functions, the testes also secrete steroid and protein hormones, a role known as the endocrine functions of the testis. The testis produces androgens such as testosterone (T), dihydrotestosterone and estrogen which are the most typical steroids, and also releases proteins called inhibins, activins, and anti-Mullerian hormone (AMH)/Mullerian-inhibiting substance (MIS) [1]. Collectively these hormones maintain the health of the testes and ensure its proper functioning regarding sperm production and delivery. In the following sessions, the androgens and the testicular protein hormones are discussed in detail.

2. The androgens

Androgens are required for the development and maintenance of specific reproductive tissues in men, such as the testis, prostate, epididymis, seminal vesicle, and penis, as well as other male characteristics such as increased muscle strength, hair growth, and so on. To maintain a sufficient androgen concentration, androgen development rates must be balanced against excretion rates and metabolic clearance [2]. The actions of the androgens are influenced by the steroid concentration that can penetrate target cells, the degree of metabolic conversion within the cells, interactions with receptor proteins, and, finally, the action of androgen receptors at the genomic level. These hormones regulate the development of the male reproductive system, as well as the development of "masculine" physical characteristics such as beards and a deep voice, as well as sexual activities [3].

It is worth noting that the testis' secretion of androgens begins even before birth. For example, the testes begin generating testosterone during the first or second part of pregnancy, depending on the species of the animal. The presence of human chorionic gonadotropin at the 7th week of intrauterine life (IUL) brings about some hormonal biosynthesis activities in the testis, leading to early expression of testosterone, which further leads to more activation of the testicular hormonal function utero. In other words, male genitalia development (and brain masculinization) require high amounts of androgens, which are produced by fetal Leydig cells [4]. However, due to the fact that placental hormones are eliminated after delivery, the pituitary-gonadal axis undergoes significant alterations, and the newborn male enters a new phase of gonadal endocrine activity which eventually leads to the death of the fetal Leydig cells. With the death of these cells, androgen production decreases, reaching a trough at postpartum. With the formation of mature Leydig cells from stem cells, testosterone levels eventually rise to high levels. Furthermore, there is also a definite sex difference in this regard, since quantitatively significant ovarian steroid production does not begin until puberty in humans. The synthesis of androgens and protein hormones such as anti-Mullerian hormone by the fetal testis, which is addressed later in this chapter, certainly plays a role in the establishment of male genital differentiation. Female differentiation, on the other hand, occurs more independently, regardless of ovarian hormone output. The androgens are very important in male reproduction [5]. Therefore, the following sessions will be devoted to explaining the physio-chemistry of the testicular androgens (testosterone, dihydrotestosterone).

3. Testosterone

This is the most common form of androgen produced by the testes. It is responsible for the growth of male genitals and sperm production. Testicles in a healthy male can produce about 6 milligrams of testosterone each day. It is synthesized and secreted by the Leydig cells of the testis. These cells do not contain 210-hydroxylase or 110-hydroxylase and so do not synthesize glucocorticoids or mineralocorticoids like the adrenal cortex, which also secrete testosterone [6]. Luteinizing hormone, in conjunction with adrenocorticotropic hormone produced by the adrenal cortex, increases testosterone synthesis by stimulating cholesterol *desmolase,* which in turn helps in steroidogenesis [7].

The prostate contains 5a-reductase which converts testosterone to its active form, dehydrotestosterone [7]. The testicular synthesis of testosterone is controlled by the activities of the hypothalamic-pituitary control mechanism. In this system, the arcuate nuclei of the hypothalamus secrete gonadotropic releasing hormone (GnRH) into the hypothalamic-hypophysial portal blood which stimulates the anterior pituitary to secrete Follicle stimulating hormone (FSH) and luteinizing hormone (LH). The luteinizing hormone is then transported to the testis where it stimulates the Leydig cells to produce testosterone [8]. This action is regulated by the presence of D-aspartic acid that is present in the pituitary gland and the testes and has a role in the regulation and release of LH and testosterone (T). In other words, luteinizing hormone acts on the Leydig cells to promote testosterone secretion, while testosterone acts as an intra-testicular paracrine mechanism to reinforce the spermatogenic effects of FSH on the Sertoli cells as well as the germ cells [9].

4. Biosynthesis of testosterone

The androgens are secreted in the testis by interstitial cells of Leydig, which account for 20% of the adult testis mass. Leydig cells are abundant in both newborn and adult males. However, in childhood, these cells are scarce or nonexistent. As a result, androgen secretion occurs in newborns and after puberty. Its secretion begins in the seventh week of fetal life by the fetal genital ridge. Around the second to fourth month of fetal life, the testes begin to secrete testosterone. Human chorionic gonadotropins, which are secreted by the placenta during pregnancy, stimulate testosterone secretion from the testes. However, until the age of 10 to 12 years old, almost no testosterone is secreted. Following that, testosterone secretion begins, and it rapidly increases at the onset of puberty and continues for the rest of one's life. After 40 years, the secretion begins to decline and reaches near zero by the age of 90 [3].

Cholesterol, a substrate that can be synthesized from acetate de novo or taken up from plasma lipoproteins, is a precursor in the synthesis of steroids at any stage. The low density lipoprotein fraction appears to be the primary extracellular store of cholesterol in human Leydig cells [10]. Furthermore, intracellular lipid droplets containing cholesterol esters can serve as intracellular cholesterol stores. The Cytochrome P_{450} side chain cleavage enzyme of the inner mitochondrial membrane of Leydig cells hydroxylates the side chains C_{22} and C_{20} of cholesterol to convert it into pregnenolone. It is then trans-located to smooth ER for conversion to testosterone in two pathways.

4.1 Dehydroepiandrosterone pathway

In this system, cholesterol is converted to pregnenolone by cAMP. Pregnenolone is then hydrolyzed to 17α-hydroxypregnenolone by the action of 17α-hydroxylase. The end product of the hydroxylation is converted in the Leydig cells to dehydroepiandrosterone (DHEA) by the actions of 17, 20-Lyase. Two events then follow. Firstly, DHEA is converted by the action of 17-HSD to androstenediol and then to testosterone by the combined action of 3-HSD and 5-4 isomerase. Secondly, DHEA is converted to androstenedione with the help of 3-HSD and 5-4 isomerase and then to testosterone by the action of 17-HSD [11].

4.2 Progesterone pathway

In this system, most of the pregnenolone is converted to progesterone in presence of 3β-HSD and 5, 4 isomerase. The activities of 17-hydroxylase convert progesterone to 17-hydroxyprogesterone, which is then converted to androstenedione by 17, 20-Lyase, and finally to testosterone by the action of 17-OHSD.

5. Actions of testosterone

The hormone helps in the differentiation of the epididymis, vas deferens, and seminal vesicle. It is also responsible for the pubertal growth spurt and the cessation of the pubertal growth spurt (epiphyseal closure). It ensures libido, spermatogenesis in Sertoli cells (paracrine effects), deepening of the voice in males, increased muscular mass, and growth of the penis and seminal vesicles, and negative feedback control of the anterior pituitary. In order to carry out its role in spermatogenesis, testosterone is said to mediate maintenance of the blood-testis barrier (BTB) [11].

6. Degradation of testosterone

Many target tissues convert testosterone into dehydrotestosterone, the most active androgen. Some tissues, including adipose tissue, the hypothalamus, and the liver, convert testosterone to estradiol. The liver degrades the majority of testosterone into inactive androsterone and dehydroepiandrosterone which are then conjugated and excreted in the urine [3].

7. Dihydrotestosterone

This metabolite of testosterone is considered the most active form of testosterone and it helps in the differentiation of the penis, scrotum, and prostate (including growth of the prostate), male hair pattern, and baldness, as well as sebaceous gland activity. It is believed that the mechanisms of action of dihydrotestosterone are the same as those of testosterone since it is the active form of testosterone.

8. Mechanism of action of testosterone and dihydrotestosterone (DHT)

Both DHT and testosterone have the ability to bind to the androgen receptor, but DHT has a higher affinity than unbound or unchanged testosterone. When DHT or testosterone binds to the androgen receptor, it forms a complex that undergoes structural changes. This complex then enters the nucleus of the cell and binds to specific nucleotide sequences of DNA known as hormone response elements. This binding causes changes in the transcription of various proteins mediated by specific genes, resulting in the androgenic effects of the cells [3].

8.1 Androgen insensitivity disorder

This is a condition characterized by the appearance of physical traits of a woman in a person who is genetically male. This disorder is caused by a deficiency of androgen receptors in target tissues of males, thereby leading to the absence of the desirable actions of testosterone and dihydrotestosterone. In this condition, female external genitalia are present but no internal genital tract. Serum testosterone levels are also very high in these individuals [12].

9. Estrogen

Testicular estrogen is formed from testosterone in Sertoli cells and it is also said to be necessary for spermeogenesis. This hormone is essential for modulating libido, erectile function, and spermatogenesis [13].

9.1 Testicular production of estrogens

Over 80% of plasma estradiol and 95% of plasma estrone in adult men is formed by extragonadal and extraadrenal aromatization of circulating testosterone and androstenedione under the control of the enzyme, aromatase. The testes produce the remaining percentage of estradiol and estrone. The Leydig cells produce larger amount of these while a minor quantity is also produced by Sertoli cells aromatizing androgens. The normal plasma estradiol levels in men range from 20 to 50 pg/mL (73–184 pmol/L), while the total production rate is around 50 g/d (184 nmol/d). These values may rise as men become older.

9.2 Mechanism of action of estrogen

Estrogen action is induced through interaction with specific nuclear estrogen receptors (ERs), which are ligand-inducible transcription factors that regulate the expression of target genes following hormone binding. There are two types of ERs: estrogen receptor (ER) and the more recently discovered estrogen receptor (ER). These two ER subtypes have distinct ligand specificities and transcriptional activity, and they mediate the classic, direct, ligand-dependent pathway involving estrogen response elements in target gene promoters and protein-protein interactions with several transcription factors. The transcriptional activity of these two ERs differs with ERα exhibiting a lower transcriptional activity due to the presence of different ERβ isoforms. Furthermore, the co-expression of both ERα and ERβ in the same cell results in a complex cross-talk that eventually results in the antagonistic effect of ER on ER-dependent transcription [14].

10. Gonadal proteins

The gonadal proteins are inhibins, activins, and anti-Mullerian hormone/ Mullerian-inhibiting substance. In the following session, I will be discussing these protein substances in relation to testicular function.

11. Inhibin

Inhibin is a dimeric glycoprotein secreted by Sertoli cells that suppresses follicle-stimulating hormone (FSH) secretion from the pituitary. Two bioactive forms of inhibin exist, inhibin A and B. Observational and experimental evidence from several studies suggest that inhibins are gonadal messengers that exert a physiological negative feedback control on FSH release at the pituitary gland. During increased rate of spermatogenesis, there is a simultaneous increase in inhibin secretion to acts on anterior pituitary and inhibits the secretion of FSH, leading to decrease in the pace of spermatogenesis [15]. Inhibin B is the circulating form of inhibin produced primarily in the testis by Sertoli cells. With the changing role of the Sertoli cell in immature and adult testes, there are temporal changes in inhibin expression and secretion. Inhibin B levels in adults are positively correlated with Sertoli cell function, sperm number, and spermatogenic status and negatively correlated with FSH. It is also important to note that a complex interaction between FSH, Sertoli cells, Leydig cells, and germ cells regulates inhibin B production. Inhibin may also play a role at an autocrine or paracrine level in modulating the actions of activin. Concerning the mechanism of action of inhibin, it is important to note that the receptors, co-receptors and intracellular signaling molecules thus far implicated in the inhibin mechanism of action are all expressed in the testis. Type II activin receptor and Mothers against decapentaplegic homolog (SMAD) proteins have been localized in Sertoli, Leydig, and germ cells, whereas the inhibin co-receptors betagly-can and inhibin-binding protein seem to be restricted to Leydig cells. However, the physiological role of paracrine/ autocrine inhibin effects within the testis has not been clarified [16].

12. Activin

Like inhibin, activin is a member of the transforming growth factor beta (TGFβ) superfamily of ligands initially identified based on their abilities [16] to augment the gonadotropin-releasing hormone (GnRH)-mediated release of FSH. The hormone is named "activin" due to its opposing effects on the functionalities of inhibin. This protein also augments erythropoietin (EPO)-dependent hemoglobin production in K562 erythroleukemia cells and enhances the proliferation of erythrocyte precursors from human bone marrow cells. Like inhibin, there are two classes of activin called activin-A and active-B [17].

Activin B from the anterior pituitary has paracrine effects on gonadotropes, enhancing GnRH-induced FSH expression and release while activin-A augments GnRH-induced LH production and is antagonized by testosterone. Activin-A has its highest concentrations in the immediate post-natal period during which it is involved in the developmental regulation of both germ cells and Sertoli cells under the modulation of follistatin [18]. Although activin-A levels are much lower in adult testes, interleukin-1 stimulates its formation in the Sertoli cell while FSH inhibits it. Due to a lack of an appropriate assay, little is known about activin-B synthesis [17].

13. Anti-Mullerian hormone (AMH)

Anti-Mullerian hormone (AMH) is a 140-kDa dimeric glycoprotein of the transforming growth factor-beta superfamily produced by the Sertoli cells of the testis. It induces regression of the Mullerian ducts during the male sex differentiation [19]. The hormone is initially synthesized as pre-prohormone, forming homodimers. The mature hormone is glycosylated and dimerized before being secreted, resulting in a 144-kDa dimer made up of identical disulphide-linked 72-kDa monomer subunits. Each monomer contains an N-terminal domain (pro region) and a C-terminal domain (mature region). The N-terminal domain enhances the activity of the C-terminal domain, which houses the molecule's bioactivity. Between 5 and 20% of AMH is cleaved during cytoplasmic transit at a particular location between the N-terminal and C-terminal domains of the 72-kDa monomer, resulting in two polypeptides of 58-kDa (pro region) and 12-kDa (mature region). These two components of the molecule are still attached non-covalently. The AMH gene is found on the short arm of chromosome 19 and has been sequenced and isolated in humans [19].

14. Mechanism of action/signal pathway of AMH

As a member of the TGFβ family of growth factors, it employs the same signal transduction mechanism as the other members of the family. These factors communicate via a serine–threonine kinase receptor complex made up of ligand-specific type II receptors and more generic type I receptors known as activin receptor-like protein kinases (ALKs) [20]. The cytoplasmic SMAD proteins are phosphorylated and activated by an active receptor complex, translocate to the nucleus and influence gene expression directly or indirectly. Anti-Mullerian hormone type II receptor (AMHRII) has been identified as being particularly necessary for AMH signaling. Furthermore, the AMH receptor gene is located on the long arm of chromosome 12. However, this gene is only expressed in the testis and neighboring mesenchymal cells to the Müllerian ducts [19].

15. Physiologic roles of AMH in males throughout life

Together with inhibin B and FSH, anti-müllerian hormone is said to be the earliest hormone secreted by the Sertoli cells in males. Therefore, the hormone is an important indicator of Sertoli cell function and it plays a vital role in the development and functions of the male reproductive system [19]. For example, this hormone is necessary for fetal sex differentiation throughout the prenatal period. Before the seventh week of pregnancy, male and female fetuses have separate gonads, bipotential external genitalia, and two pairs of unipotential internal ducts (the Müllerian ducts and the Wolffian ducts). In the XY fetus, the expression of the SRY gene, the sex reversal gene on the Y chromosome, causes testicular differentiation [19].

From the eighth week of pregnancy onwards, the developing testes' somatic cells, Leydig cells, and Sertoli cells release testosterone, Insulin-like factor-3 and AMH necessary for the individual's normal male differentiation. Although testosterone has been discussed earlier, it is important to state here that it promotes the differentiation of the Wolffian ducts into seminal vesicles, vasa deferentia, and the epididymis during fetal development. Insulin-like factor-3, also produced by Leydig cells, acts as an important signal during the first phase of testicular descent. The regression of the müllerian ducts leading to differentiation of the genitalia into differentiated into the oviducts, uterus, and upper portion of the vaginal canal in the female fetus is due to the action of AMH [21].

The infantile GnRH surge causes a significant increase in gonadotropins (LH and FSH) in the first few weeks of life, followed by an increase in levels of testosterone, AMH and inhibin B. AMH levels remain high throughout the prepubertal period of life and are reduced during puberty as testosterone levels rise. During prepubertal periods, Leydig cells produce low amounts of testosterone and Sertoli cells are still immature, making the process of formation of spermatozoa arrested in a premeiotic stage while AMH remains secreted at a high level till the onset of puberty. Leydig cells in men undergo additional differentiation at the start of puberty following the secretion of GnRH and the consequent "LH surge" [19].

There is also an increase in testosterone biosynthesis, which leads to the maturation of Sertoli cells. Sertoli cell maturation causes germ cells to undergo meiosis, which starts the process of sperm formation. The inhibitory action of testosterone triumphs over FSH stimulation, resulting in a decrease in AMH expression and the consequent decrease in its circulating levels. In adult males, AMH secretion reaches a plateau and remains nearly constant for the rest of a man's life [19].

16. Control of endocrine functions of the testis

The endocrine functions of the testis are controlled by the hypothalamus and the pituitary gland. The hypothalamus secrets gonadotropin releasing and inhibitory substances that regulate the release of the gonadotropins from the anterior pituitary. Gonadotropins, in turn, regulate the hormone-producing activities of the testis vise vasa [2].

17. Conclusion

The testes (male gonads) secrete the steroid and protein hormones. While the testicular steroid hormones are testosterone, dihydrotestosterone and estrogen, the protein hormones are inhibins, activins and anti-Mullerian hormone (Mullerian-inhibiting substance). Collectively, these hormones maintain the health of the testes, ensure proper production of sperm cells, and control the entry of the sperm cells into the female reproductive tract during ejaculation.

Objectives of the chapter

By the end of this chapter, the reader should be able to:

i. Identify the hormone produced by the testes

ii. Describe the mechanisms of action of testosterone, estrogen, inhibin, activin and a*nti-Mullerian hormone*

iii. State the roles of the various testicular hormones on male reproduction

iv. Define androgen insensitivity disorder

v. Enumerate the physiological roles of a*nti-Mullerian hormone* in males.

Author details

Emojevwe Victor[1*], Igiehon Osarugue[1], Oyovwi Mega Obukohwo[2], Nwangwa Eze Kingsley[3] and Naiho Alexander Obidike[3]

1 Department of Physiology, University of Medical Sciences, Ondo City, Ondo State, Nigeria

2 Department of Human Physiology, Achievers University, Owo, Ondo State, Nigeria

3 Department of Physiology, Delta State University, Abraka, Delta State, Nigeria

*Address all correspondence to: vemojevwe@unimed.edu.ng

IntechOpen

© 2021 The Author(s). Licensee IntechOpen. This chapter is distributed under the terms of the Creative Commons Attribution License (http://creativecommons.org/licenses/by/3.0), which permits unrestricted use, distribution, and reproduction in any medium, provided the original work is properly cited.

References

[1] Barrett KE, Barman SM, Boitano S, Brooks HL. Ganong's Review of Medical Physiology. 25th ed. New York: McGraw Hill; 2016. pp. 418-422

[2] Oyovwi MO, Nwangwa EK, Ben-Azu B, Rotue RA, Edesiri TP, Emojevwe V, et al. Prevention and reversal of chlorpromazine induced testicular dysfunction in rats by synergistic testicle-active flavonoids, taurine and coenzyme-10. Reproductive Toxicology (Elmsford, N.Y.). 2021;**101**: 50-62. DOI: 10.1016/j.reprotox.2021.01.013

[3] Sembulingam K, Sembulingam P. Essentials of Medical Physiology. 7th ed. London: Jaypee Brothers Medical Publishers; 2018. pp. 473-491

[4] Ilacqua A, Francomano D, Aversa A. The physiology of the testis. In: Belfiore A, LeRoith D, editors. Principles of Endocrinology and Hormone Action. Endocrinology. Cham: Springer; 2018. DOI: 10.1007/978-3-319-44675-2_17

[5] Browne P, Place NJ, Vidal JD, Moore IT, Cunha GR, Glickman SE, et al. Endocrine differentiation of fetal ovaries and testes of the spotted hyena (Crocuta crocuta): Timing of androgen-independent versus androgen-driven genital development. Reproduction (Cambridge, England). 2006;**132**(4): 649-659. DOI: 10.1530/rep.1.01120

[6] Kalfa N, Gaspari L, Ollivier M, Philibert P, Bergougnoux A, Paris F, et al. Molecular genetics of hypospadias and cryptorchidism recent developments. Clinical Genetics. 2019;**95**(1):122-131. DOI: 10.1111/cge.13432

[7] Goldenberg L, So A, Fleshner N, Rendon R, Drachenberg D, Elhilali M. The role of 5-alpha reductase inhibitors in prostate pathophysiology: Is there an additional advantage to inhibition of type 1 isoenzyme? Canadian Urological Association Journal. 2009;**3**(3 Suppl 2): S109-S114. DOI: 10.5489/cuaj.1114

[8] Padmanabhan V, Cardoso RC. Neuroendocrine, autocrine, and paracrine control of follicle-stimulating hormone secretion. Molecular and Cellular Endocrinology. 2020;**500**: 110632. DOI: 10.1016/j.mce.2019.110632

[9] Di Fiore MM, Boni R, Santillo A, Falvo S, Gallo A, Esposito S, et al. D-aspartic acid in vertebrate reproduction: Animal models and experimental designs‡. Biomolecules. 2019;**9**(9):445. DOI: 10.3390/biom9090445

[10] Prince FP. The human Leydig cell. In: Payne AH, Hardy MP, editors. The Leydig Cell in Health and Disease. Contemporary Endocrinology. Totowa: Humana Press; 2007. DOI: 10.1007/978-1-59745-453-7_5

[11] Schiffer L, Barnard L, Baranowski ES, Gilligan LC, Taylor AE, Arlt W, et al. Human steroid biosynthesis, metabolism and excretion are differentially reflected by serum and urine steroid metabolomes: A comprehensive review. The Journal of Steroid Biochemistry and Molecular Biology. 2019;**194**:105439. DOI: 10.1016/j.jsbmb.2019.105439

[12] Melo KF, Mendonca BB, Billerbeck AE, et al. Clinical, hormonal, behavioral, and genetic characteristics of androgen insensitivity syndrome in a Brazilian cohort: Five novel mutations in the androgen receptor gene. The Journal of Clinical Endocrinology and Metabolism. 2003;**88**(7):3241-3250. DOI: 10.1210/jc.2002-021658

[13] Yaşar P, Ayaz G, User SD, Güpür G, Muyan M. Molecular mechanism of estrogen-estrogen receptor signaling.

Reproductive Medicine and Biology. 2016;**16**(1):4-20. DOI: 10.1002/rmb2.12006

[14] Horstman AM, Dillon EL, Urban RJ, Sheffield-Moore M. The role of androgens and estrogens on healthy aging and longevity. The Journals of Gerontology. Series A, Biological Sciences and Medical Sciences. 2012;**67**(11):1140-1152. DOI: 10.1093/gerona/gls068

[15] O'Connor AE, De Kretser DM. Inhibins in normal male physiology. Seminars in Reproductive Medicine. 2004;**22**(3):177-185. DOI: 10.1055/s-2004-831893

[16] Nandedkar TD. Testicular hormones. In: Kumar A, Sharma M, editors. Basics of Human Andrology. Singapore: Springer; 2017. DOI: 10.1007/978-981-10-3695-8_8\

[17] Bristol-Gould SK, Kreeger PK, Selkirk CG, Kilen SM, Cook RW, Kipp JL, et al. Postnatal regulation of germ cells by activin: The establishment of the initial follicle pool. Developmental Biology. 2006;**298**(1):132-148. DOI: 10.1016/j.ydbio.2006.06.025

[18] Namwanje M, Brown CW. Activins and inhibins: Roles in development, physiology, and disease. Cold Spring Harbor Perspectives in Biology. 2016;**8**(7):a021881. DOI: 10.1101/cshperspect.a021881

[19] Zec I, Tislaric-Medenjak D, Megla ZB, Kucak I. Anti-Müllerian hormone: A unique biochemical marker of gonadal development and fertility in humans. Biochemia Medica. 2011;**21**(3):219-230. DOI: 10.11613/bm.2011.031

[20] Hata A, Chen YG. TGF-β signaling from receptors to Smads. Cold Spring Harbor Perspectives in Biology. 2016;**8**(9):a022061. DOI: 10.1101/cshperspect.a022061

[21] Rey RA, Musse M, Venara M, Chemes HE. Ontogeny of the androgen receptor expression in the fetal and postnatal testis: Its relevance on Sertoli cell maturation and the onset of adult spermatogenesis. Microscopy Research and Technique. 2009;**72**(11):787-795. DOI: 10.1002/jemt.20754

Chapter 3

Seminiferous Tubules and Spermatogenesis

Amor Houda, Shelko Nyaz, Bakry Mohamed Sobhy, Almandouh Hussein Bosilah, Micu Romeo, Jankowski Peter Michael and Hammadeh Mohamad Eid

Abstract

One of the major concerns of the world health community is the infertility. The definition of infertility according to the World Health Organization (WHO) and the American Society for Reproductive Medicine (ASRM) is the inability of a healthy couple to achieve a conception after one year of regular, unprotected intercourse. Fertility complications affect seven percent of the male. The causes of infertility were divided to non-obstructive and obstructive. But, in almost 75% of male infertility cases are idiopathic with predominance of the genetic abnormalities. Numerical or structural chromosomal abnormalities are considered as genetic abnormalities that occur during the meiotic division in spermatogenesis. These abnormalities get transferred to the Offspring, which affects the normal and even the artificial conception. In the human reproduction, sperm cells are considered as a delivery vehicle for the male genetic material packed in chromosomes, which are composed of nearly 2-meter Deoxyribonucleic acid (DNA) molecule and their packaging proteins. This chapter points to grant a summarized description of individual components of the male reproductive system: the seminiferous tubule and spermatogenesis. Here, we describe step by step the structure of the testis seminiferous tubule and what occurs inside these tubules like cell communication and germ cell development from spermatogonia until spermatozoon. This book chapter is very useful for the biologists and physicians working in Assisted reproduction field to understand the physiology and pathology of spermatogenesis.

Keywords: Seminiferous tubules, Spermatogenesis, Chromatin remodeling

1. Introduction

Testes or testicles appear as a pair of oval-shaped complex organs enclosed in the scrotum and based behind the penis and in front of the anus. They produce male reproductive cells, spermatozoa, and androgens, the male hormones [1]. Each adult testis weights 12 to 19 g, 4.5x 2.5x 3 cm in dimension and is suspended in the scrotum by a spermatic cord. The rete testis at the mediastinum of the testis connects to the head of epididymis, which is opposed to the testis posteriorly [2].

The tunica albuginea, fibrous capsule, covers each testis. The testis is divided by partitions of the tissue from the tunica albuginea into approximately 250 lobes. Three to ten coiled tubules are inside each lobe. These tubules are called seminiferous tubules containing two different cells population: spermatogenic or germ cells and Sertoli cells surrounded by peritubular myoid cell layer. The stroma between the seminiferous tubules is called the interstitium (interstitial tissue), where located blood and lymphatic vessels, the steroidogenic Leydig cells and other cell types [3, 4].

1.1 Tunica albuginea

Each testis is enclosed in a thick fibrous envelope, formed by collagen fibers impregnated with elastic fibers (5% elastin), called tunica albuginea. Besides, it is formed by two layers: outer longitudinal layer and inner circular layer [5].

Because of her contractile properties (erection), the tunica albuginea has different physiological functions: the preservation of the interstitial pressure inside the testis, the support of the spermatozoa movement from the testis to the epididymis, and the regulation of the blood movement through the testis. On the posterior surface of the testis, the tunica albuginea become thicker to form the mediastinum testis from which Septula testis enter the gland, separating it into almost 250 testicular lobules [6].

1.2 Basement membrane

The basement membrane is a fibrous matrix formed by type IV collagen, glycoproteins and lamin produced by the epithelial cells. It plays a crucial part in keeping up the structural and functional integrity of tissues in the testis [7, 8].

Modified cellular layer structure has been related with extreme function abnormalities of the testis like cryptorchidism, autoimmune orchitis, vasectomy [9].

1.3 Peritubular cells

In human testicular, the outer coat of the seminiferous tubules is formed by several layers of myoid, peritubular cells and extracellular matrix (ECM) proteins. The cells are peritubular myofibroblast-like cells that encompass the seminiferous tubules to maintain its structural integrity and are capable of tubular contractility and sperm transport [10].

These cells, in adults, express markers for smooth-muscle-like cells similar to the smooth muscle actin [11].

Immunohistochemical studies suggested that the cellular phenotypes differ between the outer and inner layers. After stain, the inner layers showed a smooth muscle phenotype after stain with desmin. While the outer layers were stained with vimentin indicating a connective tissue phenotype [12].

A basal lamina separates the spermatogonial stem cells (SSC) and the peritubular myoid cells (PMC). This can indicate a possible cellular interaction between the PMCs and SSC to maintain the SSCs niche, similarly to Sertoli cells [4]. One of the contribution mechanisms is through the production of secreted factors like glial cell line-derived neurotrophic factor (GDNF), that acts in combination with the androgen receptor (AR) [13].

It has also been demonstrated that the PMCs produce colony-stimulating factor 1 (CSF-1) in interaction with specific receptor CSF-1R regulate the SSCs activity [4, 14].

1.4 Leydig cells

Leydig cells called also interstitial cells because they are in the stroma between the seminiferous tubules: the interstitium (interstitial tissue) holding the tubules together within each lobule. This tissue is activated at puberty through the interstitial-cell-stimulating hormone of the anterior lobe of the pituitary gland [15].

After stimulation through the luteinizing hormone (LH), the production of testosterone via the Leydig cells increases through the regulation of the expression level of steroidogenic enzymes like the 17-β hydroxysteroid dehydrogenase [16].

Testosterone exerts its effects locally by binding to the androgen receptor (AR) within the testis or distantly by binding to androgen binding protein (ABP) which increases its levels in the seminiferous tubules and its carrying to the epididymis [17].

Elevated levels of serum LH, as well as FSH and lowered levels of serum testosterone, suggested Leydig and germ cell failure [18].

2. Structure and function of the seminiferous tubules

The seminiferous tubules are the basic units of the testicles where the SSCs proliferate and differentiate through cyclic events (mitosis, meiosis, postmeiotic spermatid development, and spermiogenesis) to generate spermatozoa in a process called spermatogenesis [19].

In humans, the seminiferous tubules represent about 60% of the total testicle volume and they are about 200 μm in diameter and have a total length of ~600 meters. These seminiferous tubules are composed of the lamina propria (peritubular tissue) with about 80 μm height and the germinal epithelium with about 8 μm thickness [20].

The germinal epithelium composed of large sertoli cells and spermatogonial germ cells. These cells are connected via tight junctions [20].

The stage of the seminiferous epithelium cycle has influence on the architecture of seminiferous tubule sections. In addition, the nerves, lymph vessels and blood vessels do not penetrate the seminiferous tubule and are located only on interstitial tissue [21].

The seminiferous tubules have terminal ends in the mediastinum testis and evacuate via straight tubular extensions called Tubuli seminiferi rect [22].

2.1 Sertoli cells

Enrico Sertoli was the author of the first publication reporting the existence of Seroli cells [23]. Later, numerous reviews in scientific journal and books have been published describing the Sertoli cell morphology and functions, mostly focusing on mammals [24].

In humans, Sertoli cells are crucial for testis physiology [25]. They proliferate during the perinatal and neonatal period, becoming quiescent for several years and having a second peak of proliferation just before puberty [26, 27].

Although, around puberty Sertoli cells stop proliferating and start to differentiate, being therefore able to support full spermatogenesis. The establishment of the Sertoli cell barrier and fluid secretion/lumen formation are clear character of Sertoli cells maturation [23, 25, 28].

Follicle stimulating hormone (FSH) and androgens are considered important factors that regulate Sertoli cell proliferation [25, 29]. In addition, oestrogens, activins, TGF-beta, BMPs, interleukins and TNF- alpha are factors involved by proliferation and differentiation of Sertoli cells [30, 31].

Sertoli cells were identified as 'nurse cells' because they are morphologically reshaped by the developing germ cells and have multitude cytoplasmatic processes. Each Sertoli cell is "nursing" approximately 30–50 germ cells at four or five diverse stages of their advancement at any given time throughout the epithelial cycle [32, 33].

Structural characteristics of the Sertoli cells varies among species, such as the heavily vacuolated nucleolus present in some ruminants [34], the nucleus localization in the middle of the seminiferous epithelium in monkeys, the presence of Charcot-Bottcher cristaloids in men [23], and the presence and amount of lipid droplets and glycogen in the Sertoli cell cytoplasm [35].

Therefore, the Sertoli cell shape may vary according to the species and the progression of spermatogenesis and the tasks. As the germ cell requirements changed, interactions and metabolic needs change substantially and accordingly, high variations are detected on the Sertoli cell cytoplasm extension, the number of nuclear pores, the presence and translocation of organelles and the protein expression pattern and location across the different phases of spermatogenesis [36, 37].

On the other hand, the Sertoli cells are considered as "epithelial" cells as they are based on a strikingly thick basal lamina, appear a remarkable design (polar-basolateral-apical) with horizontal cell–cell intersections and border on a luminal space [38].

Although Sertoli cells extend from the basement membrane of the seminiferous tubule into the adluminal compartment, the two tubular compartments are isolated by tight and adherent junction complexes between neighboring Sertoli cells, that works as the major component of the blood-testis barrier (BTB). These junctions generate the required chemical environment for fulfillment of meiosis and spermiogenesis [39].

Besides, the molecular character of Sertoli cells changed from keratin IFs to vimentin IFs during their development and maturation. Also, a wealth of special and rather extended forms of adherents junctions connected Sertoli cells and spermatogenic cells instead of the typical epithelial junctions [38].

Functionally, they play a critical role during the spermatozoa development by supporting and organizing spermatogonial germ cells during different stages of spermatogenesis through secretion of androgen-binding protein and interaction with Leydig cells [23, 28]. In addition, they provide the germ cells with a variety of ions, nutriments, carbohydrates, hormones, and growth factors [40, 41].

2.2 The transition region

The seminiferous tubules connect to the rete testis in a region named: Transition region. This region might be a specific area for immature Sertoli cells [27, 42, 43]. Also, transitional region contain a subpopulation of mitotically active Sertoli cells without differentiation, Sertoli cells markers like transcription factor GATA-4 and the androgen receptor [27, 42–44]. It can be assumed that adult Sertoli cells population is not morphologically homogeneous. As the transition region presents modified Sertoli cells that exhibit features that resemble undifferentiated Sertoli cells, with less indentations, smaller nucleolus, and more peripheral heterochromatin [45, 46]. Therefore, the dogma that the adult Sertoli cells population constitutes a terminally differentiated population in mammals has been challenged by several recent studies [23, 42, 47, 48].

In addition, because this transitional region of mammalian testis also contains spermatogonial stem cells, it has been supposed that the transition region might be an area where the seminiferous tubules continue to grow in sexually mature individuals [27, 42, 44]. Other indicated that the transition region is a site where seminiferous tubules are originally formed [49].

2.3 Spermatogonial stem cell's niche

The spermatogonial stem cell's niche (SSCs) microenvironment has a complex regulation that involves the vascular network, macrophages, the basement membrane, peritubular myoid cells, and Sertoli cells. Meantime, the stimulation of SSCs differentiation involves Leydig cells [4, 50–53].

The number of Sertoli cells per testis determines the number of available spermatogonial stem cell niches and, consequently, reflects the magnitude of sperm production capacity. Therefore, this Sertoli cell regulation ensures a proper gem cell homeostasis and regulates the germ cell density observed in the seminiferous epithelium [54].

Depending on the stimulus, a balance between differentiation and self-renewal factors regulates the fate of SSCs that are capable of self-renewal, differentiation and/or entering apoptosis [54].

In mammals, recent studies have demonstrated that the transition region is the closed niche area. The Sertoli cells in this region produce high amount of glial cell-line derived neutrophic factor (GDNF), maintaining the neighboring spermatogonia in an undifferentiated state [44].

Also, the Sertoli cells play a key role in the functional regulation of spermatogonial stem cells niche, where other somatic testicular cells, extracellular matrices and soluble factors actively participate in the complex interaction/signaling with these spermatogonial cells [4].

Several studies have demonstrated that SSCs are usually located in the seminiferous tubules area facing blood vessels of the testis interstitial compartment. It is speculated that FSH, coming from the blood vessels, stimulates Glial cell derived neurotrophic factor (GDNF) synthesis of surrounding Sertoli cells [55].

GDNF and fibroblast growth factor 2 (FGF2) are considered as the most important factors for the regulation of SSCs niche. The GDNF drive SSCs to self-renewal by binding glial cell line derived neurotrophic factor family receptor alpha 1 (GFRA1), a membrane receptor located at the surface of undifferentiated spermatogonia [10, 56, 57].

The secretion of GDNF is cyclic and coincident with the differentiation of SSCs to type A spermatogonia that are committed to spermatogenesis (density-dependent regulation), the lowest values of this peptide are found in last development stage near spermiation area [58].

Other important factors produced by Sertoli cells are leukemia inhibitory factor (LIF) and wingless-related MMTV integration site 5A (WNT5A), essential peptides that promote spermatogonial stem cell survival [4, 23].

2.4 Sertoli cell efficiency/spermatogenic efficiency

The key qualitative and quantitative determinants of sperm production are firstly the total number of Sertoli cells in the testicles and secondly their proper interactions with spermatogonial germ cells and the total number of these cells per Sertoli cells (Sertoli cell efficiency) [23, 28].

Sertoli cells show distinct capacities to hold germ cell development that varies among species. Each Sertoli cell can support a relatively fixed, species-specific, number of germ cells.

For instance, whereas chinchilla Sertoli cell can support 14 spermatids, each human Sertoli cell is able to support only 3 spermatids, resulting respectively in a huge difference in sperm production per testis gram per day (~60 vs. 4–4.5 million) between these species [1, 23, 25].

However, spermatogenic efficiency continually reduces and this characteristic is highly associated with the Sertoli cell support capacity, which decreases from around 100–150 (in fish) to 3 (in humans) spermatids for each Sertoli cell [23].

The Sertoli cells size and the space that they occupy in the seminiferous epithelium is another important factor to be considered. Species with reduced Sertoli cells occupancy in the seminiferous epithelium like mice for example (~15%), present higher spermatogenic efficiencies when compared to humans (~40%) [1].

Furthermore, the spermatogenic cycle lengths controlled by the germ cell genotype play a crucial role in determining the efficiency of spermatogenesis [1, 59]. The faster the cell differentiation process from spermatogonia to spermatozoa, the higher the daily sperm output is. If the spermatogenic cycle takes about 9 to 12 days, then the total duration of spermatogenesis (that takes almost ~4.5 cycles) will be 40 to 54 days. Spermatogenesis, in humans, takes a quite long duration (~70 days) [1].

However, germ cell loss particularly in mammalian, which is quite frequent during the spermatogonial and meiotic phases of spermatogenesis (DNA damage), influences significantly the total sperm production [60–63]. Therefore, Sertoli cell efficiency is critical in deciding the frequency of sperm production [1].

2.5 Relationship between germ cells and Sertoli cells

Spermatozoa production and maintenance throughout life is very complex and the fine regulation of spermatogenesis is under tight control and regulation [23, 28, 64–66].

Therefore, interactions among testicular cells, specially between germ cells and Sertoli cells, are crucial to preserve and regulate spermatogenesis in a very coordinated and organized manner, providing all the necessary structural and nutritional support for the developing germ cells. These interactions are important to ensure the development and completion of spermatogenesis [23, 25, 67].

At their different areas/regions, the Sertoli cells present the following contacts and functions with germ cells. In the basal compartment of the seminiferous epithelium, Sertoli cells regulate spermatogonia self-renew and differentiation [68], create contact with spermatocytes on its lateral side and, regulate the meiotic process from the duplication of DNA to the formation of spermatids [69]. In addition, in the adluminal/apical portions, Sertoli cells interact specifically with spermatids, regulate their morphology, controlling spermiation and reabsorb the residual bodies [70].

The germ cells are attached by desmosome-like, ectoplasmic specializations to Sertoli cells. On its basal side Sertoli cells contact spermatogonia through adherens junctions (AJ), guiding their homing, niche, and colonization [24]. At their adluminal aspect, Sertoli cells contact elongated spermatids through ectoplasmatic specialization, organizing the movement of these haploid cells as well as their release during spermiation [24, 71].

Germ cells like spermatocytes and early spermatids are attached by desmosomes and gap junctions. One of the most studied constitutive proteins of the gap junction is connexin 43 [72].

Also, the expression of connexin 43 differs according to the stage of germ cell development, suggesting that a particular group of germ cells can modulate this protein expression in somatic cells [73]. In humans, connexin 43 is observed in Sertoli, spermatogonia and spermatocytes cells, which suggests an accurate communication among these cells [74].

Domke et al. found that the Sertoli cells relate to each other and with the germ cells by special N-cadherin-based AJ-type junctions (Bareae adherents), which in many regions are characterized by cytoplasmic AJ plaques containing proteins p120 and p0071, plakoglobin, and α- and β-catenin [38].

2.6 Seminiferous epithelial cycle, stages, and wave

The cycle of the seminiferous epithelium refers to all the cellular interactions that occur between Sertoli cells and developing germ cells within the seminiferous epithelium in both adluminal and basal compartment.

Consequently, various cycles of spermatogenesis coincide within the seminiferous epithelium at each given time. The duration of each cycle is 16 days [75]. In humans, the cycle happens in segments rather than simultaneously around the entire periphery of the seminiferous tubule as occurs in some animals (**Figure 1**).

The stages of the cycle of seminiferous epithelium can be differentiated based on meiotic divisions, the arrangement of spermatids in the germinal epithelium and the modification of the spermatid nucleus shape and based on the development of spermatids morphology including the acrosome [77, 78].

The arrangement of stages in the seminiferous tubule of a man is different from that in most other species. The patch work of stages is arranged in a helical pattern, so a single cross section contains cells from more than one stage. Many studies have highlighted the importance of using the seminiferous epithelium cycle (SEC) stages to better understand the kinetics of the seminiferous epithelium (SE) [68, 70].

In most animals there is a wave of spermatogenesis going in an orderly fashion down the seminiferous tubule. In the human, however, there is a mosaic arrangement of the six stages of spermatogenesis [79].

These stages are designed by Roman numerals. Each of these stages has a characteristic spermatid development step. Six stages were identified in cross sections of the seminiferous tubules and were described in the spermatid maturation step and termed as follow: Sa-1 and Sa-2, Sb-1 and Sb-2, Sc-1 and Sc-2. Each one of these stages might be characterized with morphological characteristics (**Figure 1**) [80].

The spermatogenic wave is a spatial organization along the length of the tubule occurring at a single moment in time. Morphological examination of cross-sections of seminiferous tubules reveals six typical cellular associations based on the type

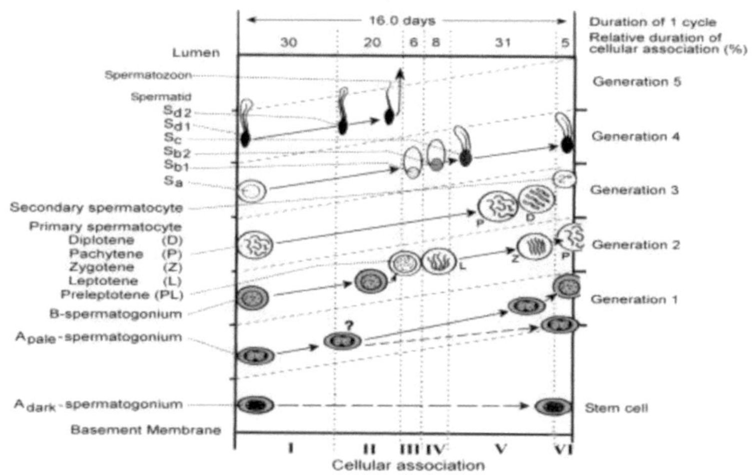

Figure 1.
The cycle of the seminiferous epithelium in humans (adapted from Amann [76]).

and stage of germ cells present in a given segment. A wave contains all 14 stages of the cycle, as well as any modulations that may be present in that segment of the tubule [76].

3. The blood testicle barrier/Sertoli cell-seminiferous epithelium barrier and spermatogenesis

Chiquoine was the first to describe the blood-testis barrier (BTB) or the Sertoli cell seminiferous epithelium barrier [81]. BTB is considered as one of the tightest blood-tissue barriers in the human body. It subdivides the seminiferous tubules epithelium into two compartments: the basal compartment and the apical compartment (**Figure 2**) [71, 82].

BTB is constituted almost exclusively by an inter-Sertoli cell junctional complex located near the basement membrane of the seminiferous tubule's epithelium. Behind the BTB localized the adluminal compartment which is a particular microenvironment that is significantly different from the interstitial space and the systemic circulation [7].

Many researchers have demonstrated that the BTB is established by actin-based tight junction (TJ), gap junction, intermediate filament-based desmosome, as well as basal ectoplasmic specialization (ES) [83–85].

However, different signaling molecules and signal pathways are controlling the BTB functions [83, 86]. Across the seminiferous epithelium in the testis, cellular events are tightly coordinated as shown by various researchers who demonstrate the existence of a local autocrine-based regulatory axis to coordinate these events. During spermiation this axis coordinates the release of spermatozoa at the apical

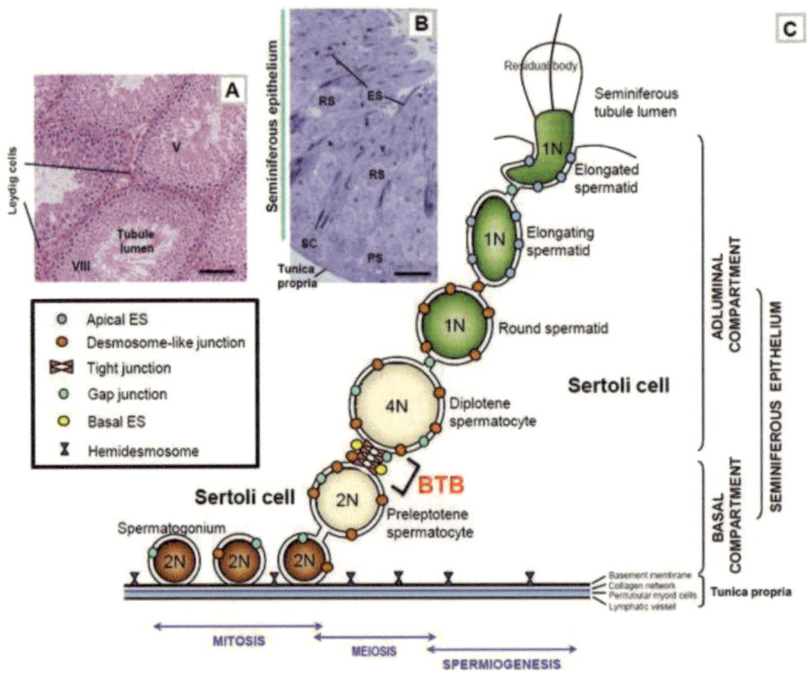

Figure 2.
The position of the BTB in the seminiferous tubules epithelium (B) of an adult mammalian testis (A) and its physiological relationship with developing germ cells during spermatogenesis (PS, pachytene spermatocyte; RS, round spermatid; ES, elongating spermatid) that are tightly associated with the Sertoli cells (SC) (adapted from Cheng and Mruk [82]).

ES at the luminal edge near the tubule lumen in the adluminal compartment at late stage III, which coincide with the remodeling of the basal ES/BTB to promote and support import of preleptotene spermatocytes that raised in the basal compartment traversing the immunological barrier at stage VIII of the epithelial cycle [1, 21, 87].

There are different types of distinct adherent Junctions (AJs) between the Sertoli cells and spermatogonial cells in the basal part of the Sertoli-Sertoli cells (heterotypic-basolateral junctions), and between Sertoli-Sertoli cells (homotypic), and between the adluminal pockets of the Sertoli cells and the spermatid heads (heterotypic-apical junctions). Therefore, Sertoli cell barrier considered one of tightest barrier in mammals [71].

Tight junctions (TJs) are the main component of the BTB that are found between adjacent Sertoli cells. They divided the seminiferous epithelium into basal compartment harbor spermatogonia and young spermatocytes and adluminal compartments where spermatocytes and spermatids are located [25, 71, 88].

Gap junctions, desmosomes, and two types of adherents junctions, testis-specific (tubulobulbar complexes and ectoplasmic specialization are other components of the sertoli cell barrier [25, 71, 88].

Spermatogenesis takes place stepwise in various segment of the seminiferous epithelium and is associated with extensive Adherent Junctions (AJs) restructuring between Sertoli cells and spermatogonial cells in the basal part of the Sertoli cells (heterotypic-basolateral junctions), and between Sertoli-Sertoli cells (homotypic), as well as between the adluminal pockets of the Sertoli cells and the spermatid heads (heterotypic-apical junctions) [41].

Spermatogonial renewal, differentiation, and cell cycle progression up to the preleptotene spermatocyte stage developed in the basal compartment of the epithelium outside the BTB.

These BTB undergoes reconstruction to allow the transit of preleptotene spermatocytes connected by intercellular bridges as clones at stage VIII of the seminiferous epithelial cycle of spermatogenesis [87, 89].

Therefore, preleptotene spermatocytes are the germ cells that pass the BTB as clones linked by intercellular bridges, that will differentiate into spermatocytes (zygotene and diplotene), to be pursued by two meiotic divisions (meiosis I and II) to form haploid spermatids in the apical compartment behind the BTB [89].

Meiosis I and II, spermiogenesis, and spermiation all take place in a specialized microenvironment in the adluminal compartment behind the BTB [71].

Therefore, the passage across the Sertoli cell barrier is a remarkable achievement, because the spermatocytes are no single cells, but form syncytia in which the cells are connected through cytoplasmic bridges [90].

Smith and Braun have provided critical insights of the molecular mechanism underlying this process. They revealed that an intermediate compartment enclosing the migrating spermatocytes was formed by "new" and "old" TJs above and below the spermatocytes, respectively [91].

Claudin-3 was transiently incorporated into the new tight junctions (TJs) and then replaced by claudin-11. Dissolution of the old TJs released the spermatocytes into the adluminal compartment. Also, when the syncytium moves toward the adluminal compartment, the BTB is opened on the adluminal side and is simultaneously closed on the basal side, at all to prevent the barrier from becoming leaky [91].

Therefore, BTB barrier created to protect germ cells undergoing meiosis from autoimmune reaction. The blood testes barrier was identified as a major barrier between the germinal epithelium and the interstitium of the testis. Also, this barrier established an immune privileged environmental with the seminiferous tubules [23, 42, 92].

By dividing the seminiferous epithelium, the BTB selectively inhibits the passage of many substances included in the general circulation [39, 93].

However, as a barrier, TJs restrict free passage of water, solutes, and ions. As a fence, TJs divide the plasma membrane into basolateral and apical regions, which confers cell polarity [94–96].

Nevertheless, during spermiogenesis, less than 25% of haploid spermatids become spermatozoa and at spermiation, could be liberated into the tubule lumen. The other portion of spermatids undergoes apoptosis [61, 97].

4. Spermatogenesis

Spermatogenesis is a very well-organized temporal process. It included different steps leading to a chronological evolution from totipotent, primitive stem cells (spermatogonial stem cells, SSCs) to a spermatogonium transformed to a specialized cell: Spermatozoon. Also, spermatogenesis is a highly organized process in which the germ cells go through several divisions and intricate differentiation steps, resulting in the production of the spermatozoa (**Figure 3**) [80].

4.1 Germ cell migration and development

Germ cell development is a lengthy and complex process that starts with specification in the early embryo and proceeds through stages of migration, proliferation, epigenetic reprogramming, sex differentiation, and gametogenesis to ultimately produce mature oocyte or sperm [98].

Therefore, Germ cells have a pivotal role in development by transmitting genetic information to the next generation. During germ cell development, epigenetic marks are erased and subsequently re-established during gametogenesis [99].

Embryonic bipotent primordial germ cells (PGCs) are among the first lineages established in early embryonic development, and the successful passage of these dedicated precursors from their birthplace to the developing gonad (gonad primordia) ensures an adequate supply of gametes for reproduction in the adult [100–102].

In the embryo, PGCs migrate from the proximal epiblast to the gonadal ridge, where they are enclosed by Sertoli and peritubular myoid cells, forming the seminiferous cords. In the developing seminiferous cords, the PGCs become gonocytes,

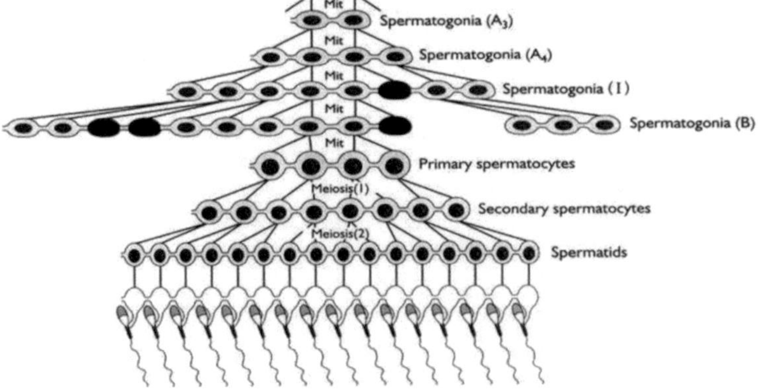

Figure 3.
Different steps of spermatogenesis leading from spermatogonia to spermatozoon.

which home to the basement membrane of the seminiferous tubules, where they differentiate into spermatogonial stem cells (SSCs) and initiate self-renewing divisions [103].

In other words, Primordial germ cells (PGC) take up residence at the testicular tubules during embryonic development, and undergo mitotic proliferation and differentiation, to become the spermatogonial stem cells which stand at the basis of spermatogenesis, starting at the onset of puberty [103]. By the end of PGC migration, around 5% of migratory cells remain outside the gonad and later undergo apoptosis [104].

During their migration however, heterogeneity of cellular behavior is observed due to change in cellular morphology from the time of specification to colonization [100].

SSCs, precursors originate from the PGCs, are usually located in a distinct position inside the seminiferous epithelium, referred to as the spermatogonial stem cell niche [105, 106].

SSCs within their niche either self-renew, remain quiescent, or generate spermatogonia committed to differentiation [4, 51, 107, 108].

Immature (fetal/neonatal) SSC precursors are commonly referred to as gonocytes or prospermatogonia, which are considered quiescent from the time of colonizing the seminiferous cords. This quiescence continues until they re-enter the cell cycle, migrate to the basement membrane, and undergo maturation and differentiation, either to constitute the SSC pool or differentiate into spermatogonia that will later become sperm [109–111].

Therefore, spermatogonial stem cells (SSCs) are the basis of spermatogenesis and maintain the continuous sperm production required for male fertility [107, 112].

More than a decade is required for prepubertal testis development and gonadal maturation in humans, generally characterized by the existence of a juvenile pause and an extended time span of prepubertal development [113–115].

Testicular tissue reactivation at the puberty, called gonadarche (earliest gonadal changes), occurs between 9 and 13 years in human. Before gonardache, there is a period of gonadal dormancy, characterized by low gonadotropin secretion, minimal testosterone secretion, discontinued Sertoli cell proliferation, and variable mitotic activity of germ cells human [116–118].

In humans, the neonatal maturation of the testis in mammals is commonly characterized by an early testosterone peak. The testosterone peak occurs after several months [119, 120].

It is associated with the movement of gonocytes to the basement membrane. Hence, this migration toward the basement membrane can take up to nine months [121]. Some studies have reported some spermatogonial heterogeneity neonatally and the appearance of differentiating spermatogonia prior to puberty [103].

4.2 Spermatogonial stage

The first phase in spermatogenesis is the proliferation and differentiation of spermatogonia.

The fetal spermatogonia develop first into transitional spermatogonia and then to spermatogonial stem cells forming the spermatogonial stem cell niche and located in the basal compartment of seminiferous tubule. These are classified into three categories: dark type (A_{dark}), pale type (A_{pale}) and type B spermatogonia [55].

Type A_{dark} spermatogonia reproduce via mitosis to generate both dark and pale spermatogonia. Throughout adult live, undifferentiated A_{pale} Spermatogonia (A_{pale}) periodically divide, giving rise to B spermatogonia (B). The A_{dark} spermatogonia (A_{dark}) are quiescent reserve cells [80].

Another proliferative spermatogonia type include type A isolated spermatogonia ($A_{isolated}$) which divided to form the type A_{paired} spermatogonia (A_{paired}). After 4 mitotic divisions, 16 cells of type A aligned spermatogonia ($A_{aligned}$) are formed will differentiate into A_1 spermatogonia to be followed by 6 mitotic divisions to form 1024 preleptotene spermatocytes [1].

However, some of type A spermatogonia transform to differentiated type A spermatogonia (A1, A2, A3, A4), intermediate spermatogonia (In) and then type B spermatogonia (B) (**Figure 3**) [1, 112].

It is important to mention that during the different stages of spermatogenesis, the spermatogonia remain connected by intercellular bridges to ensure the synchronization of the germ cell maturation and the biochemical interactions [122].

Besides, it is not yet clear if all cells within the type A spermatogonia pool are true spermatogonial stem cells because different studies have found counts that were significantly lower than the ones originally disclosed. Approximately 1/12–1/15 of the pool appears to be composed of true spermatogonial stem cells [51, 123].

4.3 Spermatocytogenesis

Spermatocytogenesis phase include the meiotic phase, in which primary spermatocyte in the basal compartment undergo meiosis I and meiosis II to give rise to haploid spermatids that are released from the seminiferous epithelium at spermiation area [77, 124].

Mitosis involves the proliferation and maintenance of spermatogonia. The mitotic phase involves spermatogonia (types A and B) and primary spermatocytes (spermatocytes I). Primary spermatocytes are produced by developing germ cells interconnected by intracellular bridges through a series of mitotic divisions. The tight junction barrier supports an early spermatocyte within the basal compartment and all subsequent germ cells within the adluminal compartment.

Type B spermatogonia undergo mitosis to produce primary spermatocytes, secondary spermatocytes, and spermatids. At stage VII of the epithelial cycle, type B spermatogonia differentiate into preleptotene, followed by leptotene spermatocytes, which are the primary diploid spermatocytes that cross the BTB while differentiating into zygotene spermatocytes at stages VIII-IX [125, 126].

Once in the adluminal compartment, spermatocytes undergo two consecutive rounds of meiosis at stage XIV. During the first meiotic division (reduction division), the primary spermatocytes divided to form secondary spermatocytes. The spermatocyte needs almost 26 days to be mature. Spermatocytes type I undergoes a long prophase in the first division, therefore they have the longest life span. The prophase of the second meiotic division is very short, thus secondary spermatocytes have a short life span [80].

4.4 Spermiogenesis

To obtain a hydrodynamic sperm head and to protect the paternal genome from any modifications during his journey through the male and female reproductive tracts, the human sperm DNA, in early spermiogenic phases, undergoes major cellular and nuclear changes [97].

Spermiogenesis is the process of differentiation of the spermatids into spermatozoa with fully compacted chromatin. A process of metamorphosis occurs from a round cell with typical organelles to a highly specialized, elongated cell. Later, the spermatid undergoes a series of morphological changes (Head, midpiece and tail) and their chromatin structure and function change [97].

The spermatid undergoes the Golgi phase, which is marked by the formation of the polarity. The head is at one end covered by Golgi apparatus developed later to acrosome and contain the synthesized proteolytic enzyme. The midpiece is at the other end, in which the mitochondria accumulates and one of the centriole pair elongates to form the tail or flagellum [97].

The post meiotic phase, progressive condensation of the chromatin occurs with inactivation of the genome. The meiotic phase involves primary spermatocytes until spermatids are formed, and during this process, chromosome pairing, crossover, and genetic exchange take place until a new genome is determined. Meiosis consists of two successive divisions to yield four haploid spermatids from one diploid primary spermatocyte. After the first meiotic division (reduction division), each daughter cell contains one partner of the homologous chromosome pair, and they are called secondary spermatocytes (2n). These spermatids are haploid with (22, X) or (22, Y) chromosome and undergo complete differentiation/morphogenesis known as Spermiogenesis [127].

During this process, morphological changes, chromatin structural and functional modifications occur once the process of meiosis is completed (**Figure 4**).

4.5 Chromatin remodeling during spermatogenesis

Sperm cells are remarkably different from somatic cells in their chromatin structure. During spermatogenesis, the majority of histones replace transition proteins and protamine (small highly basic proteins bound to the sperm DNA) [128].

So, in early spermiogenic phases, a major chromatin packaging takes place. The nucleosome-bound DNA configuration will be first destabilized by hyperacetylation of the canonical histones, which will neutralize the positive charge of lysine, reducing their affinity for DNA and by the DNA topoisomerase II (topo II), which will cause double and single DNA strands breaks to reduce the tension of the DNA [129].

The chromatin in the elongated nucleus became ten times more compacted than the chromatin in the nucleus of a somatic cell through progressive modifications (**Figure 5**) [130].

Post-translational modifications of the proteins facilitated the transition histone-protamine: Acetylation, ubiquitination, and phosphorylation of histone H4, phosphorylation and dephosphorylation of the transition proteins. (Adapted from Braun [130]).

Sperm chromatin is tightly packaged by protamines, while up to 15% of the histones remain in the mature human spermatozoa [131], these retained histones within the sperm nucleus possibly have a contribution on sperm function [132].

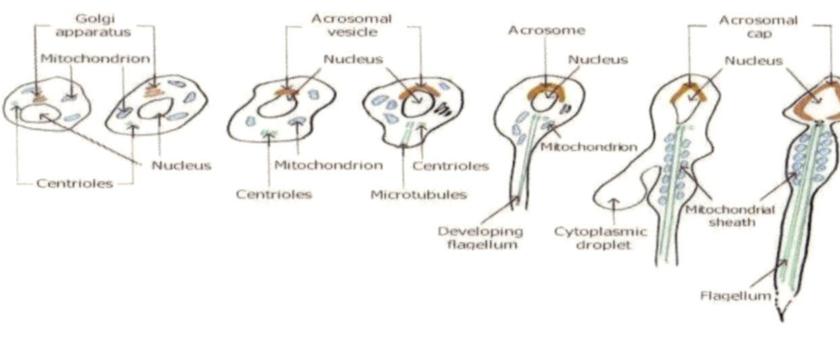

Figure 4.
Developmental changes in the spermatid [127].

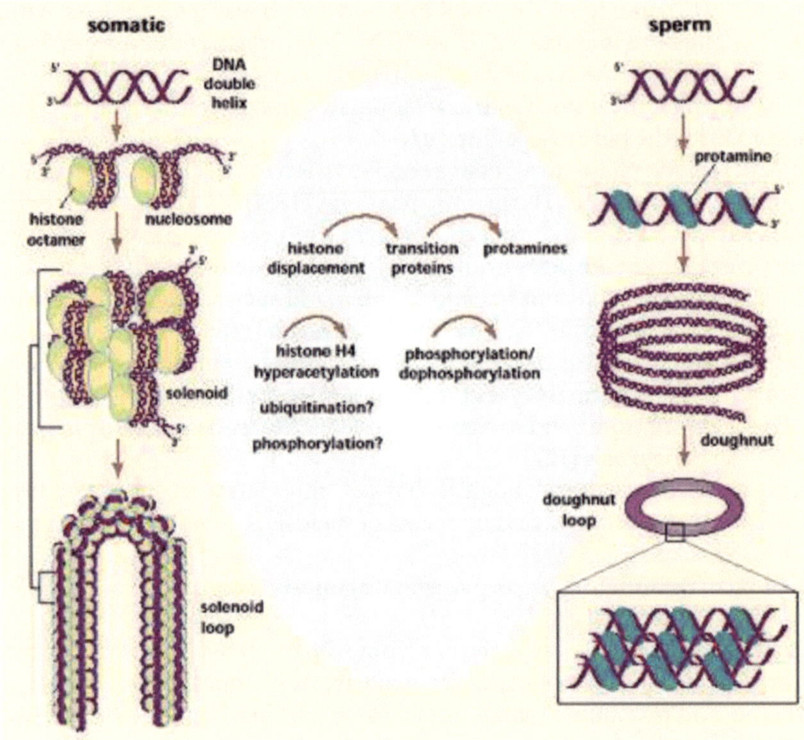

Figure 5.
A representation of the difference in the chromatin packaging between a somatic cell (left) and a sperm (right). The chromatin converted from a nucleo-histone structure (solenoid loop) into nucleo-protamine structure (toroid: Doughnut loop).

It concerns mainly the transcription sites of genes in sperms that are important for the preservation of the paternal genome epigenetics for their later expression during the early embryonic development [133]. The regulatory sequences [134], microRNA clusters, Transcription factors, paternally imprinted genes [132], the centromeric and telomeric DNA [135], retroposons [136], matrix associated regions [137], genes that produces rRNA, are transcribed at the final stages of spermatogenesis [138].

In fact, there are imprinted genes in the male genome, epigenetic changes in the DNA and nucleoproteins that edit the chromatin to make it ready for the control of the embryonic growth and development [139] and step by step the chromatin will be genetically silenced in the spermatozoa [140].

Also, after spermatocytogenesis, the chromatin structural changes will be more obvious when two smaller more basic proteins (10–20% lysine and arginine) named "Transition proteins" TP1 and TP2 are synthesized and deposed at the mid-stage of spermatids formation (**Figure 5**).

TP2 (13KDa) appears in step 1 and TP1 (6,2KDa) appears in step3. At this time, most of the core histones are eliminated, and the chromatin structure becomes more condensed. As their name indicate these proteins stay only for a short period of time attached to the DNA [141].

The transition proteins are then replaced by sperm-specific nuclear basic proteins (protamines), which are a synthesis in the last spermatid stage and play a vital role in the condensation and stabilization of sperm chromatin [142].

Humans sperm contain two protamines, protamine 1 (P1) and protamine 2 (P2), both are expressed in roughly similar quantities with a mean P1/P2 ratio of approximately 1.0 [143].

The sperm protamine 1 (P1) (51 AA) is the first to be synthesized as a mature protein [144]. The protamine 2 (P2) is formed as a precursor which is twice size as P1 (101 residues) and undergoes cleavage by proteolysis after its deposition onto sperm DNA to eliminate a short fragment of peptide [145]. Nanassy et al. suggested a clinical value of the protamine ratio between 0.54 and 1.43 for a fertile, normozoospermic man [146]. Any change in the ratio P1/P2 or between histone and protamines in the human sperm will be associated with the low compaction of human chromatin, which results in DNA fragmentation, lower fertilization rates, and reduced pregnancy rates [143, 147, 148]. Finally, a mature spermatid frees itself from the Sertoli cells and in a process called spermiation and enters the seminiferous tubule as a spermatozoon.

4.6 Hormone regulation of spermatogenesis

Spermatogenesis is controlled through several hormones. The first control is through a neurological pathway; the gonadotrophin-releasing hormone (GnRH) secreted by the hypothalamus stimulate the adenohypophysis to secret the luteinizing hormone (LH) and the follicle-stimulating hormone (FSH). The LH stimulates the Leydig cells to produce the Testosterone, and the FSH assists the Sertoli cells to support the spermatozoa during the different phases of spermatogenesis. Beside FSH and LH there are other hormones which plays crucial roles during spermatogenesis like the prolactin, and the growth hormone (**Figure 6**) [80].

Besides, anti-Mullerian hormone (AMH), which promotes the regression of Müller's ducts as the male foetus develops produced by Sertoli cells [149, 150]. In addition, inhibin and activin secreted by Sertoli cells, Activin increases the FSH levels needed for semen production. Whereas Inhibin regulate FSH secretion by the hypothalamus and helps maintain testicular homeostasis [149]. Sertoli cells only syndrome is characterized by the exclusive presence of Sertoli cells (without germ cells) in seminiferous tubules, making spermatogenesis impossible [151].

Furthermore, Leydig and Sertoli cells produce reproductive steroid hormones. Leydig cells secrete several different types of androgens, including dihydrotestosterone and testosterone, which modulate the development and maturation of spermatozoa [152].

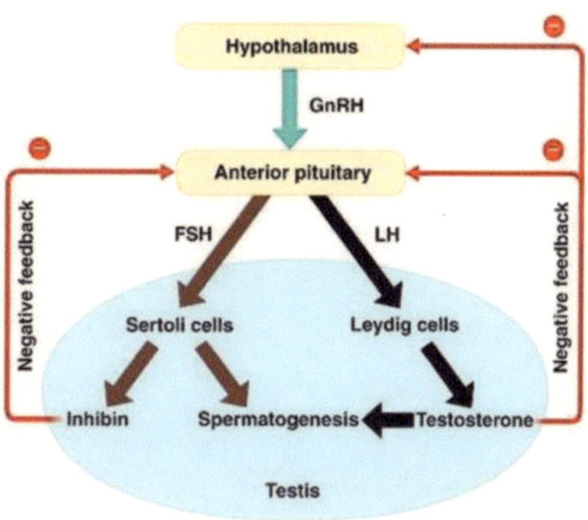

Figure 6.
Hormonal regulation of spermatogenesis.

Author details

Amor Houda[1], Shelko Nyaz[2], Bakry Mohamed Sobhy[3], Almandouh Hussein Bosilah[3], Micu Romeo[4], Jankowski Peter Michael[1] and Hammadeh Mohamad Eid[1*]

1 Department of Obstetrics and Gynaecology, Saarland University, Germany

2 Community Health Department, Technical College of Health, Sulaimani Polytechnic University, and Medical Laboratory Science, College of Health Science, University of Human Development, Kurdustan, Iraq

3 Obstetrics and Gynecology, Medical Faculty, Fayoum University, Egypt

4 Obstetrics and Gynecology Department, University of Medicine and Pharmacy, Cluj-Napoca, Romania

*Address all correspondence to: mehammadeh@yahoo.de; mohamad.eid.hammadeh@uks.eu

IntechOpen

© 2021 The Author(s). Licensee IntechOpen. This chapter is distributed under the terms of the Creative Commons Attribution License (http://creativecommons.org/licenses/by/3.0), which permits unrestricted use, distribution, and reproduction in any medium, provided the original work is properly cited.

References

[1] Hess RA, De Franca LR. Spermatogenesis and cycle of the seminiferous epithelium. Adv Exp Med Biol [Internet]. 2008 [cited 2021 May 15];636:1-15. Available from: https://link.springer.com/chapter/10.1007/978-0-387-09597-4_1

[2] Robaire B, Hinton BT. The Epididymis. In: Knobil and Neill's Physiology of Reproduction: Two-Volume Set. Elsevier Inc.; 2015. p. 691-771.

[3] Kemal H, Gülkesen T, Erdoêru C, Figen S!, Karpuzoêlu G. Expression of extracellular matrix proteins and vimentin in testes of azoospermic man: an immunohistochemical and morphometric study.

[4] Oatley JM, Brinster RL. The Germline Stem Cell Niche Unit in Mammalian Testes. Physiol Rev [Internet]. 2012 [cited 2021 May 15];92:577-95. Available from: www.prv.org

[5] Davis JR, Langford GA. Response of the testicular capsule to acetylcholine and noradrenaline [25] [Internet]. Vol. 222, Nature. Nature; 1969 [cited 2021 May 13]. p. 386-7. Available from: https://pubmed.ncbi.nlm.nih.gov/5782120/

[6] Setchell BP, Davies R V, Gladwell RT, Hinton BT, Main SJ, Pilsworth L, Waites GMH. The movement of fluid in the seminiferous tubules and rete testis.

[7] Siu MKY, Cheng Y, Cheng CY. M i n i rev i ew Extracellular Matrix: Recent Advances on Its Role in Junction Dynamics in the Seminiferous Epithelium During Spermatogenesis 1. Biol Reprod. 2004;71:375-91.

[8] Harvey SJ, Perry J, Zheng K, Chen D, Sado Y, Jefferson B, Ninomiya Y, Jacobs R, Hudson BG, Thorner PS. Sequential expression of type IV collagen networks: Testis as a model and relevance to spermatogenesis. Am J Pathol. 2006 May 1;168(5):1587-97.

[9] Richardson LL, Kleinman HK, Dym M. Altered Basement Membrane Synthesis in the Testis After Tissue Injury. Vol. 19, Journal of Androtogy.

[10] Potter SJ, Defalco T. PROOF ONLY REPRODUCTION Role of the testis interstitial compartment in spermatogonial stem cell function. 2017; Available from: www.reproduction-online.org

[11] Albrecht M. Insights into the nature of human testicular peritubular cells. Vol. 191, Annals of Anatomy. Urban & Fischer; 2009. p. 532-40.

[12] Davidoff MS, Breucker H, Holstein AF, Seidl K. Cellular architecture of the lamina propria of human seminiferous tubules. Cell Tissue Res [Internet]. 1990 Nov [cited 2021 May 13];262(2):253-61. Available from: https://link.springer.com/article/10.1007/BF00309880

[13] Chen LY, Willis WD, Eddy EM. Targeting the Gdnf Gene in peritubular myoid cells disrupts undifferentiated spermatogonial cell development. Proc Natl Acad Sci U S A [Internet]. 2016 Feb 16 [cited 2021 May 12];113(7):1829-34. Available from: https://pubmed.ncbi.nlm.nih.gov/26831079/

[14] DeFalco T, Potter SJ, Williams A V., Waller B, Kan MJ, Capel B. Macrophages Contribute to the Spermatogonial Niche in the Adult Testis. Cell Rep [Internet]. 2015 Aug 18 [cited 2021 May 15];12(7):1107-19. Available from: https://pubmed.ncbi.nlm.nih.gov/26257171/

[15] Jones RE, Lopez KH. Human Reproductive Biology. Fourth Edi. Academic Press; 2014.

[16] Prante BC, Garman KL, Sims BN, Lindsey JS. Matrix-coated transwell-cultured TM4 sertoli cell testosterone-regulated gene expression mimics in vivo expression. Vitr Cell Dev Biol

- Anim [Internet]. 2009 Dec 23 [cited 2021 May 15];44(10):434-43. Available from: https://link.springer.com/article/10.1007/s11626-008-9135-8

[17] Smith LB, Walker WH. The regulation of spermatogenesis by androgens [Internet]. Vol. 30, Seminars in Cell and Developmental Biology. Elsevier Ltd; 2014 [cited 2021 May 16]. p. 2-13. Available from: https://pubmed.ncbi.nlm.nih.gov/24598768/

[18] Ishida H, Isurugi K, Aso Y, Takayasu H, Tamaoki BI. Endocrine studies in Sertoli cell only syndrome. J Urol. 1976 Jul 1;116(1):56-8.

[19] Schlatt S, Ehmcke J. Regulation of spermatogenesis: An evolutionary biologist's perspective [Internet]. Vol. 29, Seminars in Cell and Developmental Biology. Elsevier Ltd; 2014 [cited 2021 May 15]. p. 2-16. Available from: https://pubmed.ncbi.nlm.nih.gov/24685618/

[20] Holstein AF. Human spermatogenesis: Basic research and clinical issues. Ann Anat. 1999 Sep 1;181(5):427-36.

[21] Cheng CY, Mruk DD. A local autocrine axis in the testes that regulates spermatogenesis [Internet]. Vol. 6, Nature Reviews Endocrinology. Nature Publishing Group; 2010 [cited 2021 May 12]. p. 380-95. Available from: https://www.nature.com/articles/nrendo.2010.71

[22] de Kretser DM, Loveland K, O'Bryan M. Spermatogenesis. In: Endocrinology: Adult and Pediatric. Elsevier Inc.; 2015. p. 2325-2353.e9.

[23] França LR, Hess RA, Dufour JM, Hofmann MC, Griswold MD. The Sertoli cell: One hundred fifty years of beauty and plasticity. Andrology [Internet]. 2016 Mar 1 [cited 2021 May 15];4(2):189-212. Available from: https://pubmed.ncbi.nlm.nih.gov/26846984/

[24] Lara N de L e. M, Costa GMJ, Figueiredo AFA, de França LR. The Sertoli cell: What can we learn from different vertebrate models? In: Animal Reproduction [Internet]. Brazilian College of Animal Reproduction; 2018 [cited 2021 May 15]. p. 81-92. Available from: http://www.ncbi.nlm.nih.gov/pubmed/33299481

[25] Martins Lara NL, Avelar GF, Costa GMJ, Santos Nassif Lacerda SM, Hess RA, França LR. Cell-cell interactions-structural. In: Encyclopedia of Reproduction. Elsevier; 2018. p. 68-75.

[26] Sharpe RM, McKinnell C, Kivlin C, Fisher JS. Proliferation and functional maturation of Sertoli cells, and their relevance to disorders of testis function in adulthood. Vol. 125, Reproduction. Journals of Reproduction and Fertility Ltd; 2003. p. 769-84.

[27] Tarulli GA, Stanton PG, Loveland KL, Meyts ER-D, McLachlan RI, Meachem SJ. A survey of Sertoli cell differentiation in men after gonadotropin suppression and in testicular cancer. Spermatogenesis. 2013 Jan;3(1):e24014.

[28] Griswold MD. 50 years of spermatogenesis: Sertoli cells and their interactions with germ cells [Internet]. Vol. 99, Biology of Reproduction. Oxford University Press; 2018 [cited 2021 May 15]. p. 87-100. Available from: https://academic.oup.com/biolreprod/article/99/1/87/4862466

[29] Skinner MK, Griswold MD. Sertoli Cell Biology. Sertoli Cell Biology. Elsevier Inc.; 2005.

[30] Puglisi R, Montanari M, Chiarella P, Stefanini M, Boitani C. Regulatory role of BMP2 and BMP7 in spermatogonia and Sertoli cell proliferation in the immature mouse. Eur J Endocrinol [Internet]. 2004 Oct [cited 2021 May 17];151(4):511-20. Available from: https://pubmed.ncbi.nlm.nih.gov/15476453/

[31] Lucas TF, Nascimento AR, Pisolato R, Pimenta MT, Lazari MFM,

Porto CS. Receptors and signaling pathways involved in proliferation and differentiation of Sertoli cells. Spermatogenesis [Internet]. 2014 Jan [cited 2021 May 17];4(1):e28138. Available from: https://pubmed.ncbi.nlm.nih.gov/25225624/

[32] Kelly CW, Janecki A, Steinberger A, Russell LD. Structural characteristics of immature rat sertoli cells in vivo and in vitro. Am J Anat [Internet]. 1991 Oct 1 [cited 2021 May 17];192(2):183-93. Available from: https://onlinelibrary.wiley.com/doi/full/10.1002/aja.1001920207

[33] Risley MS, Tan IP, Farrell J. Gap junctions with varied permeability properties establish cell-type specific communication pathways in the rat seminiferous epithelium. Biol Reprod [Internet]. 2002 Sep 1 [cited 2021 May 17];67(3):945-52. Available from: http://www.biolreprod.org

[34] Steger K, Wrobel K-H. Anatomy and Erhbryology Immunohistochemical demonstration of cytoskeletal proteins in the ovine testis during postnatal development. Vol. 189, Anat Embryol. 1994.

[35] Erkan M, Sousa M. Fine structural study of the spermatogenic cycle in Pitar rudis and *Chamelea gallina* (Mollusca, Bivalvia, Veneridae). Tissue Cell. 2002 Aug 1;34(4):262-72.

[36] Johnston DS, Wright WW, DiCandeloro P, Wilson E, Kopf GS, Jelinsky SA. Stage-specific gene expression is a fundamental characteristic of rat spermatogenic cells and Sertoli cells. Proc Natl Acad Sci U S A [Internet]. 2008 Jun 17 [cited 2021 May 17];105(24):8315-20. Available from: www.pnas.org/cgi/content/full/

[37] Wright WW. Stage-specific gene expression by Sertoli cells. In: Sertoli Cell Biology. Elsevier; 2015. p. 273-306.

[38] Domke LM, Rickelt S, Dörflinger Y, Kuhn C, Winter-Simanowski S, Zimbelmann R, Rosin-Arbesfeld R, Heid H, Franke WW, Franke WW, Domke LM, Kuhn C, Rosin-Arbesfeld R. The cell-cell junctions of mammalian testes: I. The adhering junctions of the seminiferous epithelium represent special differentiation structures. Tissue Res. 2014;357:645-65.

[39] Li MWM, Xia W, Mruk DD, Wang CQF, Yan HHN, Siu MKY, Lui WY, Lee WM, Cheng CY. Tumor necrosis factor α reversibly disrupts the blood-testis barrier and impairs Sertoli-germ cell adhesion in the seminiferous epithelium of adult rat testes. J Endocrinol [Internet]. 2006 Aug [cited 2021 May 15];190(2):313-29. Available from: www.endocrinology-journals.org

[40] Eskild W, Trøen G, Blaner WS, Nilsson A, Hansson V. Evidence for independent control at the mRNA and protein levels of cellular retinol binding protein 1 in rat Sertoli cells. J Reprod Fertil. 2000;119(1):101-9.

[41] Mruk DD, Cheng CY. Sertoli-Sertoli and Sertoli-Germ Cell Interactions and Their Significance in Germ Cell Movement in the Seminiferous Epithelium during Spermatogenesis. 2004; Available from: http://www.endo-society.org

[42] Figueiredo AFA, França LR, Hess RA, Costa GMJ. Sertoli cells are capable of proliferation into adulthood in the transition region between the seminiferous tubules and the rete testis in Wistar rats. Cell Cycle [Internet]. 2016 Sep 16 [cited 2021 May 15];15(18):2486-96. Available from: https://www.tandfonline.com/action/journalInformation?journalCode=kccy20

[43] Kulibin AY, Malolina EA. Only a small population of adult Sertoli cells actively proliferates in culture. Reproduction [Internet]. 2016 Oct 1 [cited 2021 May 15];152(4):271-81. Available from: https://europepmc.org/article/med/27512121

[44] Aiyama Y, Tsunekawa N, Kishi K, Kawasumi M, Suzuki H, Kanai-Azuma M, Kurohmaru M, Kanai Y. A Niche for GFRα1-Positive Spermatogonia in the Terminal Segments of the Seminiferous Tubules in Hamster Testes. Stem Cells [Internet]. 2015 Sep 1 [cited 2021 May 12];33(9):2811-24. Available from: http://www.emdmillipore.com

[45] Osman DI, Plöen L. The Terminal Segment of the Seminiferous Tubules and the Blood-Testis Barrier Before and After Efferent Ductule Ligation in the Rat. Int J Androl. 1978;1(1-6):235-49.

[46] Nykänen M. Fine structure of the transitional zone of the rat seminiferous tubule. Cell Tissue Res [Internet]. 1979 May [cited 2021 May 17];198(3):441-54. Available from: https://link.springer.com/article/10.1007/BF00234189

[47] Hayrabedyan S, Todorova K, Pashova S, Mollova M, Fernández N. Sertoli Cell Quiescence - New Insights [Internet]. Vol. 68, American Journal of Reproductive Immunology. John Wiley & Sons, Ltd; 2012 [cited 2021 May 17]. p. 451-5. Available from: https://onlinelibrary.wiley.com/doi/full/10.1111/j.1600-0897.2012.01137.x

[48] Haverfield JT, Stanton PG, Meachem SJ. Adult Sertoli cell differentiation status in humans. In: Sertoli Cell Biology. Elsevier; 2015. p. 409-36.

[49] Malolina EA, Kulibin AY. Rete testis and the adjacent seminiferous tubules during postembryonic development in mice. Russ J Dev Biol [Internet]. 2017 Nov 1 [cited 2021 May 15];48(6):385-92. Available from: https://link.springer.com/article/10.1134/S1062360417060029

[50] Phillips BT, Gassei K, Orwig KE. Spermatogonial stem cell regulation and spermatogenesis [Internet]. Vol. 365, Philosophical Transactions of the Royal Society B: Biological Sciences. Royal Society; 2010 [cited 2021 May 17]. p. 1663-78. Available from: https://royalsocietypublishing.org/

[51] Yoshida S, Sukeno M, Nabeshima YI. A vasculature-associated niche for undifferentiated spermatogonia in the mouse testis. Science (80-) [Internet]. 2007 Sep 21 [cited 2021 May 16]; 317(5845):1722-6. Available from: http://science.sciencemag.org/

[52] Caires K, Broady J, Mclean D. REVIEW Maintaining the male germline: regulation of spermatogonial stem cells. J Endocrinol [Internet]. 2010;133-45. Available from: www.endocrinology-journals.org

[53] Heinrich A, DeFalco T. Essential roles of interstitial cells in testicular development and function [Internet]. Vol. 8, Andrology. Blackwell Publishing Ltd; 2020 [cited 2021 May 15]. p. 903-14. Available from: https://onlinelibrary.wiley.com/doi/full/10.1111/andr.12703

[54] Oatley MJ, Racicot KE, Oatley JM. Sertoli Cells Dictate Spermatogonial Stem Cell Niches in the Mouse Testis 1. Biol Reprod [Internet]. 2011;84:639-45. Available from: http://www.biolreprod.org

[55] De Rooij DG. The spermatogonial stem cell niche [Internet]. Vol. 72, Microscopy Research and Technique. John Wiley & Sons, Ltd; 2009 [cited 2021 May 15]. p. 580-5. Available from: www.interscience.

[56] Chen SR, Liu YX. Regulation of spermatogonial stem cell self-renewal and spermatocyte meiosis by Sertoli cell signaling [Internet]. Vol. 149, Reproduction. BioScientifica Ltd.; 2015 [cited 2021 May 12]. p. R159-67. Available from: https://pubmed.ncbi.nlm.nih.gov/25504872/

[57] Chen L-Y, Willis WD, Eddy EM. Targeting the Gdnf Gene in peritubular myoid cells disrupts undifferentiated

spermatogonial cell development. Proc Natl Acad Sci U S A [Internet]. 2016 Feb 16 [cited 2021 May 17];113(7):1829-34. Available from: http://www.ncbi.nlm.nih.gov/pubmed/26831079

[58] Johnston DS, Olivas E, DiCandeloro P, Wright WW. Stage-specific changes in GDNF expression by rat sertoli cells: A possible regulator of the replication and differentiation of stem spermatogonia. Biol Reprod [Internet]. 2011 Oct 1 [cited 2021 May 15];85(4):763-9. Available from: http://www.biolreprod.org

[59] França LR, Ogawa T, Avarbock MR, Brinster RL, Russell LD. Germ cell genotype controls cell cycle during spermatogenesis in the rat. Biol Reprod [Internet]. 1998 Dec 1 [cited 2021 May 17];59(6):1371-7. Available from: https://academic.oup.com/biolreprod/article/59/6/1371/2740926

[60] Russell LD, Chiarini-Garcia H, Korsmeyer SJ, Knudson CM. Bax-dependent spermatogonia apoptosis is required for testicular development and spermatogenesis. Biol Reprod [Internet]. 2002 Apr 1 [cited 2021 May 17];66(4):950-8. Available from: http://www.biolreprod.org

[61] Shaha C, Tripathi R, Prasad Mishra D. Male germ cell apoptosis: Regulation and biology [Internet]. Vol. 365, Philosophical Transactions of the Royal Society B: Biological Sciences. Royal Society; 2010 [cited 2021 May 17]. p. 1501-15. Available from: https://royalsocietypublishing.org/doi/abs/10.1098/rstb.2009.0124

[62] Aitken RJ, Curry BJ. Redox regulation of human sperm function: From the physiological control of sperm capacitation to the etiology of infertility and DNA damage in the germ line. Antioxidants Redox Signal [Internet]. 2011 Feb 1 [cited 2021 May 12];14(3): 367-81. Available from: https://www.liebertpub.com/doi/abs/10.1089/ars.2010.3186

[63] Murphy CJ, Richburg JH. Implications of Sertoli cell induced germ cell apoptosis to testicular pathology. Spermatogenesis [Internet]. 2014 Mar 4 [cited 2021 May 15];4(2):e979110. Available from: https://doi.org/10.4161/21565562.2014.979110

[64] Wong EWP, Cheng CY. Chapter 7 Polarity Proteins and Cell-Cell Interactions in the Testis. Vol. 278, International Review of Cell and Molecular Biology. Academic Press; 2009. p. 309-53.

[65] Ramaiah M, Wilkinson MF. MicroRNAs and Sertoli cells. In: Sertoli Cell Biology. Elsevier; 2015. p. 307-32.

[66] Yang Q-E, Oatley JM. Early postnatal interactions between Sertoli and germ cells. In: Sertoli Cell Biology. Elsevier; 2015. p. 81-98.

[67] Yan Cheng C, Mruk DD. Biochemistry of Sertoli cell/germ cell junctions, germ cell transport, and spermiation in the seminiferous epithelium. In: Sertoli Cell Biology. Elsevier; 2015. p. 333-83.

[68] de Rooij DG. Proliferation and differentiation of spermatogonial stem cells [Internet]. Vol. 121, Reproduction. Journals of Reproduction and Fertility Ltd; 2001 [cited 2021 May 15]. p. 347-54. Available from: https://rep.bioscientifica.com/view/journals/rep/121/3/347.xml

[69] Russell L. Movement of spermatocytes from the basal to the adluminal compartment of the rat testis. Am J Anat [Internet]. 1977 Mar 1 [cited 2021 May 15];148(3):313-28. Available from: https://anatomypubs.onlinelibrary.wiley.com/doi/full/10.1002/aja.1001480303

[70] Meistrich ML, Hess RA. Assessment of Spermatogenesis Through Staging of Seminiferous Tubules. In Humana Press, Totowa, NJ; 2013 [cited 2021 May 15]. p. 299-307. Available from: https://link.

springer.com/protocol/10.1007/978-1-62703-038-0_27

[71] Yan Cheng C, Mruk DD. The blood-testis barrier and its implications for male contraception. Pharmacol Rev [Internet]. 2012 Jan [cited 2021 May 12];64(1):16-64. Available from: https://pubmed.ncbi.nlm.nih.gov/22039149/

[72] Kidder GM, Cyr DG. Roles of connexins in testis development and spermatogenesis. Vol. 50, Seminars in Cell and Developmental Biology. Academic Press; 2016. p. 22-30.

[73] Pointis G, Gilleron J, Carette D, Segretain D. Physiological and physiopathological aspects of connexins and communicating gap junctions in spermatogenesis [Internet]. Vol. 365, Philosophical Transactions of the Royal Society B: Biological Sciences. Royal Society; 2010 [cited 2021 May 15]. p. 1607-20. Available from: https://royalsocietypublishing.org/doi/abs/10.1098/rstb.2009.0114

[74] Defamie N, Berthaut I, Mograbi B, Chevallier D, Dadoune JP, Fénichel P, Segretain D, Pointis G. Impaired gap junction connexin43 in Sertoli cells of patients with secretory azoospermia: A marker of undifferentiated Sertoli cells. Lab Investig [Internet]. 2003 Mar 1 [cited 2021 May 15];83(3):449-56. Available from: https://pubmed.ncbi.nlm.nih.gov/12649345/

[75] HELLER, CG. Kinetics of the germinal epithelium. Recent Prog Horm Res [Internet]. 1964 [cited 2021 May 16];20:545-745. Available from: http://ci.nii.ac.jp/naid/10013446938/en/

[76] Amann RP. The cycle of the seminiferous epithelium in humans: A need to revisit? [Internet]. Vol. 29, Journal of Andrology. John Wiley & Sons, Ltd; 2008 [cited 2021 May 17]. p. 469-87. Available from: https://onlinelibrary.wiley.com/doi/full/10.2164/jandrol.107.004655

[77] RUSSELL, D. L. Form, dimensions and cytology of mammalian Sertoli cells. Sertoli Cell [Internet]. 1993 [cited 2021 May 16];1-37. Available from: http://ci.nii.ac.jp/naid/10026669432/en/

[78] França LR, Godinho CL. Testis Morphometry, Seminiferous Epithelium Cycle Length, and Daily Sperm Production in Domestic Cats (*Felis catus*). Biol Reprod [Internet]. 2003 [cited 2021 May 15];68:1554-61. Available from: http://www.biolreprod.org

[79] Silber SJ. Reproductive Infertility Microsurgery in the Male and Female. Baltimore: Williams & Wilkins; 1984. 296 p.

[80] Sharma R, Agarwal A. Spermatogenesis: An Overview. In: Sperm Chromatin. Springer New York; 2011. p. 19-44.

[81] Chiquoine AD. Observations on the early events of cadmium necrosis of the testis. Anat Rec [Internet]. 1964 May 1 [cited 2021 May 17];149(1):23-35. Available from: https://anatomypubs.onlinelibrary.wiley.com/doi/full/10.1002/ar.1091490104

[82] Cheng CY, Mruk DD. Regulation of blood-testis barrier dynamics by focal adhesion kinase (FAK): An unexpected turn of events [Internet]. Vol. 8, Cell Cycle. Taylor and Francis Inc.; 2009 [cited 2021 May 17]. p. 3493-9. Available from: https://www.tandfonline.com/action/journalInformation?journalCode=kccy20

[83] Lie PPY, Cheng CY, Mruk DD. Signalling pathways regulating the blood-testis barrier. Vol. 45, International Journal of Biochemistry and Cell Biology. Elsevier Ltd; 2013. p. 621-5.

[84] Pelletier RM. The blood-testis barrier: The junctional permeability, the proteins and the lipids. Vol. 46, Progress in Histochemistry and Cytochemistry. Urban & Fischer; 2011. p. 49-127.

[85] Stanton PG. Regulation of the blood-testis barrier [Internet]. Vol. 59, Seminars in Cell and Developmental Biology. Academic Press; 2016 [cited 2021 May 16]. p. 166-73. Available from: https://pubmed.ncbi.nlm.nih.gov/27353840/

[86] Li G, Xu A, Sim S, Priest JR, Tian X, Khan T, Quertermous T, Zhou B, Tsao PS, Quake SR, Wu SM. Transcriptomic Profiling Maps Anatomically Patterned Subpopulations among Single Embryonic Cardiac Cells. Dev Cell. 2016 Nov 21;39(4):491-507.

[87] Xiao X, Mruk DD, Wong CKC, Yan Cheng C. Germ cell transport across the seminiferous epithelium during spermatogenesis [Internet]. Vol. 29, Physiology. American Physiological Society; 2014 [cited 2021 May 16]. p. 286-98. Available from: www.physiologyonline.org

[88] Vogl AW, Young JS, Du M. New Insights into Roles of Tubulobulbar Complexes in Sperm Release and Turnover of Blood-Testis Barrier. In: International Review of Cell and Molecular Biology. Elsevier Inc.; 2013. p. 319-55.

[89] Miething A. Local desynchronization of cellular development within mammalian male germ cell clones. Ann Anat. 2010 Aug 20;192(4):247-50.

[90] KRESTER D, M. D. The cytology of the testis. Physiol Reprod [Internet]. 1994 [cited 2021 May 16]; Available from: http://ci.nii.ac.jp/naid/10020906012/en/

[91] Smith BE, Braun RE. Germ cell migration across sertoli cell tight junctions. Science (80-) [Internet]. 2012 Nov 9 [cited 2021 May 16]; 338(6108):798-802. Available from: https://pubmed.ncbi.nlm.nih.gov/22997133/

[92] Francavilla F, Barbonetti A, Francavilla S. Naturally-occurring antisperm antibodies in men: Interference with fertility and clinical implications. An update Testicular Cacer in infettive men View project. 2007 [cited 2021 May 15]; Available from: https://www.researchgate.net/publication/6346681

[93] Kato R, Maeda T, Akaike T, Tamai I. Nucleoside transport at the blood-testis barrier studied with primary-cultured sertoli cells. J Pharmacol Exp Ther [Internet]. 2005 Feb 1 [cited 2021 May 15];312(2):601-8. Available from: https://jpet.aspetjournals.org/content/312/2/601

[94] Shin K, Fogg VC, Margolis B. Tight junctions and cell polarity [Internet]. Vol. 22, Annual Review of Cell and Developmental Biology. Annu Rev Cell Dev Biol; 2006 [cited 2021 May 16]. p. 207-35. Available from: https://pubmed.ncbi.nlm.nih.gov/16771626/

[95] Anderson JM, Van Itallie CM. Tight junctions [Internet]. Vol. 18, Current Biology. Curr Biol; 2008 [cited 2021 May 12]. Available from: https://pubmed.ncbi.nlm.nih.gov/18957244/

[96] Furuse M. Molecular basis of the core structure of tight junctions. [Internet]. Vol. 2, Cold Spring Harbor perspectives in biology. Cold Spring Harb Perspect Biol; 2010 [cited 2021 May 15]. Available from: https://pubmed.ncbi.nlm.nih.gov/20182608/

[97] O'Donnell L, Nicholls PK, O'Bryan MK, McLachlan RI, Stanton PG. Spermiation. Spermatogenesis [Internet]. 2011 Jan [cited 2021 May 15];1(1):14-35. Available from: https://doi.org/10.4161/spmg.1.1.14525

[98] Ewen KA, Koopman P. Mouse germ cell development: From specification to sex determination. Vol. 323, Molecular and Cellular Endocrinology. Elsevier; 2010. p. 76-93.

[99] Sasaki H, Matsui Y. Epigenetic events in mammalian germ-cell development: Reprogramming and

beyond [Internet]. Vol. 9, Nature Reviews Genetics. Nat Rev Genet; 2008 [cited 2021 May 15]. p. 129-40. Available from: https://pubmed.ncbi.nlm.nih.gov/18197165/

[100] Richardson BE, Lehmann R. Mechanisms guiding primordial germ cell migration: strategies from different organisms. 2010; Available from: www.nature.com/reviews/molcellbio

[101] Wong T-T, Collodi P. Inducible Sterilization of Zebrafish by Disruption of Primordial Germ Cell Migration. 2013; Available from: www.plosone.org

[102] Barton LJ, LeBlanc MG, Lehmann R. Finding their way: themes in germ cell migration [Internet]. Vol. 42, Current Opinion in Cell Biology. Elsevier Ltd; 2016 [cited 2021 May 12]. p. 128-37. Available from: https://pubmed.ncbi.nlm.nih.gov/27484857/

[103] Guo J, Grow EJ, Mlcochova H, Maher GJ, Lindskog C, Nie X, Guo Y, Takei Y, Yun J, Cai L, Kim R, Carrell DT, Goriely A, Hotaling JM, Cairns BR. The adult human testis transcriptional cell atlas. Cell Res [Internet]. 2018 Dec 1 [cited 2021 May 15];28(12):1141-57. Available from: https://doi.org/10.1038/s41422-018-0099-2

[104] Cantú A V., Laird DJ. A pilgrim's progress: Seeking meaning in primordial germ cell migration. Stem Cell Res. 2017 Oct 1;24:181-7.

[105] Costa GMJ, Avelar GF, Rezende-Neto J V., Campos-Junior PHA, Lacerda SMSN, Andrade BSC, Thomé RG, Hofmann MC, Franca LR. Spermatogonial Stem Cell Markers and Niche in Equids. PLoS One [Internet]. 2012 Aug 28 [cited 2021 May 13];7(8). Available from: https://pubmed.ncbi.nlm.nih.gov/22937157/

[106] Chiarini-Garcia H, Raymer AM, Russell LD. Non-random distribution of spermatogonia in rats: Evidence of niches in the seminiferous tubules. Reproduction. 2003;126(5):669-80.

[107] De Rooij DG. The nature and dynamics of spermatogonial stem cells [Internet]. Vol. 144, Development (Cambridge). Company of Biologists Ltd; 2017 [cited 2021 May 15]. p. 3022-30. Available from: https://pubmed.ncbi.nlm.nih.gov/28851723/

[108] Hofmann MC. Gdnf signaling pathways within the mammalian spermatogonial stem cell niche. Mol Cell Endocrinol [Internet]. 2008 Jun 25 [cited 2021 May 15];288(1-2):95-103. Available from: https://pubmed.ncbi.nlm.nih.gov/18485583/

[109] Culty M. Gonocytes, the forgotten cells of the germ cell lineage [Internet]. Vol. 87, Birth Defects Research Part C - Embryo Today: Reviews. Wiley-Liss Inc.; 2009 [cited 2021 May 13]. p. 1-26. Available from: www.interscience.wiley.com

[110] Yoshida S, Sukeno M, Nakagawa T, Ohbo K, Nagamatsu G, Suda T, Nabeshima YI. The first round of mouse spermatogenesis is a distinctive program that lacks the self-renewing spermatogonia stage. Development. 2006 Apr 15;133(8):1495-505.

[111] Law NC, Oatley JM. Developmental underpinnings of spermatogonial stem cell establishment. Vol. 8, Andrology. Blackwell Publishing Ltd; 2020. p. 852-61.

[112] ROOIJ DG DE, RUSSELL LD. All You Wanted to Know About Spermatogonia but Were Afraid to Ask. J Androl [Internet]. 2000 Nov 12 [cited 2021 May 15];21(6):776-98. Available from: https://onlinelibrary.wiley.com/doi/full/10.1002/j.1939-4640.2000.tb03408.x

[113] Wu X, Schmidt JA, Avarbock MR, Tobias JW, Carlson CA, Kolon TF, Ginsberg JP, Brinster RL. Prepubertal human spermatogonia and mouse gonocytes share conserved gene expression of germline stem cell regulatory molecules. Proc Natl Acad

Sci U S A [Internet]. 2009 Dec 22 [cited 2021 May 16];106(51):21672-7. Available from: www.pnas.org/cgi/content/full/

[114] Grumbach MM. The neuroendocrinology of human puberty revisited. In: Hormone Research [Internet]. Karger Publishers; 2002 [cited 2021 May 15]. p. 2-14. Available from: www.karger.comwww.karger.com/journals/hre

[115] Lara N de L e. M, Costa GMJ, Avelar GF, Guimarães DA, França LR. Postnatal testis development in the collared peccary (*Tayassu tajacu*), with emphasis on spermatogonial stem cells markers and niche. Gen Comp Endocrinol. 2019 Mar 1;273:98-107.

[116] Masliukaite I, Hagen JM, Jahnukainen K, Stukenborg JB, Repping S, van der Veen F, van Wely M, van Pelt AMM. Establishing reference values for age-related spermatogonial quantity in prepubertal human testes: a systematic review and meta-analysis. Fertil Steril. 2016 Dec 1;106(7): 1652-1657.e2.

[117] Mäkelä JA, Koskenniemi JJ, Virtanen HE, Toppari J. Testis Development [Internet]. Vol. 40, Endocrine Reviews. Endocrine Society; 2019 [cited 2021 May 15]. p. 857-905. Available from: https://academic.oup.com/edrv

[118] Sharma S, Wistuba J, Pock T, Schlatt S, Neuhaus N. Spermatogonial stem cells: updates from specification to clinical relevance. Hum Reprod Update [Internet]. 2019;25(3):275-97. Available from: https://academic.oup.com/humupd/article/25/3/275/5366160

[119] Picut CA, Ziejewski MK, Stanislaus D. Review Article Comparative Aspects of Pre-and Postnatal Development of the Male Reproductive System. Birth Defects Res. 2018;110:190-227.

[120] Foster DL, Hileman SM. Puberty in the Sheep. In: Knobil and Neill's Physiology of Reproduction: Two-Volume Set. Elsevier Inc.; 2015. p. 1441-85.

[121] Drumond AL, Meistrich ML, Chiarini-Garcia H. Spermatogonial morphology and kinetics during testis development in mice: A high-resolution light microscopy approach. Reproduction. 2011 Jul;142(1):145-55.

[122] Dym M, Fawcett DW. The blood-testis barrier in the rat and the physiological compartmentation of the seminiferous epithelium. Biol Reprod [Internet]. 1970 Dec 1 [cited 2021 May 15];3(3):308-26. Available from: https://academic.oup.com/biolreprod/article/3/3/308/2768522

[123] Nakagawa T, Nabeshima Y ichi, Yoshida S. Functional Identification of the Actual and Potential Stem Cell Compartments in Mouse Spermatogenesis. Dev Cell. 2007 Feb 1;12(2): 195-206.

[124] Setchell BP. Sermatogenesis. In: The mammalian testis. London, UK: Paul Elek.; 1978. p. 181-232.

[125] Parvinen M. Regulation of the seminiferous epithelium. Endocr Rev [Internet]. 1982 Oct 1 [cited 2021 May 15];3(4):404-17. Available from: https://academic.oup.com/edrv/article/3/4/404/2548759

[126] Russell L. Movement of spermatocytes from the basal to the adluminal compartment of the rat testis. Am J Anat [Internet]. 1977 Mar 1 [cited 2021 May 16];148(3):313-28. Available from: https://anatomypubs.onlinelibrary.wiley.com/doi/full/10.1002/aja.1001480303

[127] Publikationen der UdS: Impact of Tobacco smoking on sperm nuclear proteins genes : H2BFWT, TNP1, TNP2, PRM1, and PRM2 and its influence on male infertility [Internet]. [cited 2021 May 17]. Available from: https://publikationen.sulb.uni-saarland.de/handle/20.500.11880/31318

[128] Balhorn R. The protamine family of sperm nuclear proteins [Internet]. Vol. 8, Genome Biology. Genome Biol; 2007 [cited 2021 May 12]. Available from: https://pubmed.ncbi.nlm.nih.gov/17903313/

[129] Laberge R-M, Boissonneault G. On the Nature and Origin of DNA Strand Breaks in Elongating Spermatids 1. Biol Reprod [Internet]. 2005 [cited 2021 May 15];73:289-96. Available from: http://www.biolreprod.org

[130] Braun RE. Packaging paternal chromosomes with protamine [Internet]. 2001 [cited 2021 May 12]. Available from: www.array.ucsd.edu

[131] Wykes SM, Krawetz SA. The structural organization of sperm chromatin. J Biol Chem. 2003 Aug 8;278(32):29471-7.

[132] Hammoud SS, Nix DA, Zhang H, Purwar J, Carrell DT, Cairns BR. Distinctive chromatin in human sperm packages genes for embryo development. Nature [Internet]. 2009 Jul 23 [cited 2021 May 15];460(7254):473-8. Available from: https://www.nature.com/articles/nature08162

[133] Ihara M, Meyer-Ficca ML, Leu NA, Rao S, Li F, Gregory BD, Zalenskaya IA, Schultz RM, Meyer RG. Paternal Poly (ADP-ribose) Metabolism Modulates Retention of Inheritable Sperm Histones and Early Embryonic Gene Expression. PLoS Genet [Internet]. 2014 [cited 2021 May 15];10(5):1004317. Available from: www.plosgenetics.org

[134] Castillo J, Amaral A, Azpiazu R, Vavouri T, Estanyol JM, Ballesca JL, Oliva R. Genomic and proteomic dissection and characterization of the human sperm chromatin. Mol Hum Reprod [Internet]. 2014 May 28 [cited 2021 May 17];20(11):1041-53. Available from: https://academic.oup.com/molehr/article/20/11/1041/2459854

[135] Zalenskaya IA, Zalensky AO. Non-random positioning of chromosomes in human sperm nuclei. Chromosom Res [Internet]. 2004 [cited 2021 May 17];12(2):163-73. Available from: https://link.springer.com/article/10.1023/B:CHRO.0000013166.04629.97

[136] Pittoggi C, Renzi L, Zaccagnini G, Cimini D, Degrassi F, Giordano R, Magnano AR, Lorenzini R, Lavia P, Spadafora C. A fraction of mouse sperm chromatin is organized in nucleosomal hypersensitive domains enriched in retroposon DNA. J Cell Sci [Internet]. 1999 Oct 15 [cited 2021 May 15];112(20): 3537-48. Available from: https://journals.biologists.com/jcs/article/112/20/3537/25843/A-fraction-of-mouse-sperm-chromatin-is-organized

[137] Ward WS. Function of sperm chromatin structural elements in fertilization and development [Internet]. Vol. 16, Molecular Human Reproduction. Oxford Academic; 2009 [cited 2021 May 16]. p. 30-6. Available from: https://academic.oup.com/molehr/article/16/1/30/1056814

[138] Sillaste G, Kaplinski L, Meier R, Jaakma Ü, Eriste E, Salumets A. The Authors ISSN 1470-1626 (paper). 2017;1741-7899. Available from: www.reproduction-online.org

[139] Canovas S, Ross PJ. Epigenetics in preimplantation mammalian development [Internet]. Vol. 86, Theriogenology. Elsevier Inc.; 2016 [cited 2021 May 12]. p. 69-79. Available from: https://pubmed.ncbi.nlm.nih.gov/27165992/

[140] Ren X, Chen X, Wang Z, Wang D. Is transcription in sperm stationary or dynamic? J Reprod Dev. 2017;63(5):439-43.

[141] Steger K, Klonisch T, Gavenis K, Drabent B, Doenecke D, Bergmann M. Expression of mRNA and protein of

nucleoproteins during human spermiogenesis. Mol Hum Reprod [Internet]. 1998 Oct 1 [cited 2021 May 17];4(10):939-45. Available from: https://academic.oup.com/molehr/article/4/10/939/1037678

[142] Balhorn R. Sperm Chromatin: An Overview. In: A Clinician's Guide to Sperm DNA and Chromatin Damage [Internet]. Springer International Publishing; 2018 [cited 2021 May 12]. p. 3-30. Available from: https://link.springer.com/chapter/10.1007/978-3-319-71815-6_1

[143] Oliva R. Protamines and male infertility. Human Reproduction Update. 2006.

[144] Queralt R, Adroer R, Oliva R, Winkfein RJ, Retief JD, Dixon GH. jouR.A o MOLECULAR [EVOLUTION Evolution of Protamine P1 Genes in Mammals. Vol. 40, J Mol Evol. 1995.

[145] Green GR, Balhorn R, Poccia DL, Hecht NB. Synthesis and processing of mammalian protamines and transition proteins. Mol Reprod Dev [Internet]. 1994 Mar 1 [cited 2021 May 15];37(3):255-63. Available from: https://onlinelibrary.wiley.com/doi/full/10.1002/mrd.1080370303

[146] Nanassy L, Liu L, Griffin J, T. Carrell D. The Clinical Utility of the Protamine 1/Protamine 2 Ratio in Sperm. Protein Pept Lett. 2012 Oct 30;18(8):772-7.

[147] García-Peiró A, Martínez-Heredia J, Oliver-Bonet M, Abad C, Amengual MJ, Navarro J, Jones C, Coward K, Gosálvez J, Benet J. Protamine 1 to protamine 2 ratio correlates with dynamic aspects of DNA fragmentation in human sperm. Fertil Steril. 2011 Jan 1;95(1):105-9.

[148] Simon L, Castillo J, Oliva R, Lewis SEM. Relationships between human sperm protamines, DNA damage and assisted reproduction outcomes. Reprod Biomed Online [Internet]. 2011 Dec [cited 2021 May 16];23(6):724-34. Available from: https://pubmed.ncbi.nlm.nih.gov/22036908/

[149] Kim E, Mobley III J, Stewart A, Moss J. Sertoli-Cell-Only Syndrome. Medscape Ref [Internet]. 2015; Available from: http://emedicine.medscape.com/article/437884-overview.

[150] Anniballo R, Brehm R, Steger K. Recognising the Sertoli- cell- only (SCO) syndrome: A case study. Andrologia [Internet]. 2011 Feb [cited 2021 May 12];43(1):78-83. Available from: https://pubmed.ncbi.nlm.nih.gov/21219389/

[151] Paulis G, Paulis L, Romano G, Concas C, Di Sarno M, Pagano R, Di Filippo A, Di Petrillo ML. Pregnancy and live birth after follicle-stimulating hormone treatment for an infertile couple including a male affected by sertoli cell-only syndrome. Res Reports Urol [Internet]. 2017 Oct 30 [cited 2021 May 15];9:203-8. Available from: /pmc/articles/PMC5669790/

[152] Umehara T, Kawashima I, Kawai T, Hoshino Y, Morohashi KI, Shima Y, Zeng W, Richards JS, Shimada M. Neuregulin 1 regulates proliferation of leydig cells to support spermatogenesis and sexual behavior in adult mice. Endocrinology [Internet]. 2016 Dec 1 [cited 2021 May 16];157(12):4899-913. Available from: https://academic.oup.com/endo/article/157/12/4899/2758443

Chapter 4

Positional Relationships among Male Reproductive Organs in Insects

Satoshi Hiroyoshi and Gadi V.P. Reddy

Abstract

The location, morphology and function of male internal reproductive organs in insects have been extensively studied, but the relative positioning of those organs is less understood. Position and morphology of the testis, vas deferens, seminal vesicle, accessory gland and ejaculatory duct determine the migration or ejaculation of sperm and other substances. In species where the testis is connected with the seminal vesicle directly or the seminal vesicle is lacking, males usually store complete sperm in the testis and thus can use them immediately for mating. In contrast, the testis of lepidopteran insects is separated from the duplex (sperm storage organ) via the vas deferens, and the sperm are not mature, requiring morphological development in the vas deferens. Here, we discuss the significance of various positional relationships of male reproductive organs and how this relates to their morphology and function with a focus on sperm.

Keywords: Testis, seminal vesicle, vas deferens, sperm reflux, sperm migration

1. Introduction

The morphology, structure and size of organs have functional significance. Insects are the most abundant of all organisms in terms of species number, resulting in female and male reproductive organs being highly diverse in their structure. While the general pattern of spermatogenesis in insects is basically similar to that in mammals, the morphology, structure, and size of sperm in insects are highly variable [1, 2].

Because available resources are usually limited, the number of sperm produced should be inversely proportional to their size [3]. Although smaller testes do not necessarily produce many sperm, sperm size is closely related to testis size. Among *Drosophila* groups, there are positive relationships between testis length and sperm length [4–6]. Moreover, sperm length correlates positively with body size in butterfly species [7]. How much sperm a female receives and stores in the spermatheca (female sperm storage organ) should be determined by how many sperm are used for fertilization during the post-mating period. Many insect species are known to over-ejaculate under laboratory conditions but it is unclear whether excessive ejaculation is common in the field.

In general, sperm size in animals including insects is not proportional to body size. *Drosophila* flies, as is well known, are tiny and thus have small testes, although

D. bifurca have large testes containing very long sperm reaching a length of 5.8 cm [8]. Therefore, the number of giant sperm that they produce is small [9]. This may be due to the size of the testes as well as other reproductive organs. Giant sperm are coiled [10], but their length and thickness must not hinder the movement within the male and female reproductive tract. It is thought that not only the cost of sperm production but also the control of sperm migration would be restricted to make the gamete larger, but in fact some *Drosophila* sperm have become giant beyond this constraint.

Sperm produced in the testes are usually stored in the seminal vesicles, if they are present. In insects, when present within the male reproductive organs, sperm are either not motile or their motility is more suppressive than when they are retained in the female reproductive organs. Therefore, males need to both store sperm in the seminal vesicles, and limit the energy costs associated with those sperm until mating. Recent studies have revealed that sperm age or die in the spermatheca [11, 12] or male reproductive organs [13], and it is interesting to see how aging affects sperm quality and survival in the seminal vesicles because the morphology of the seminal vesicles may be associated with sperm aging, as mentioned below. Furthermore, a recent molecular biology study indicates that increased expression of seminal fluid protein incorporated in spermatophore genes is correlated with increased sperm viability in the ejaculates [14].

During mating, male insects pass a spermatophore or bolus of seminal fluid to females. In many species, a spermatophore is produced by male reproductive accessory glands, whereas in lepidopteran insects, spermatophores are produced mainly from parts of the simplex (ejaculatory duct). In a nymphalid butterfly, *Polygonia c-aureum* L., the contents of the simplex are initially ejaculated followed by the contents of the duplex (male sperm storage organ) and then by the accessory glands just before the end of copulation [15]. In the Coleoptera, the accessory glands and seminal vesicles open directly into the ejaculatory duct, allowing seminal fluids in the accessory glands and sperm in the seminal vesicles to ejaculate simultaneously. Molecules transferred from males to females via the seminal vesicles that originated from the accessory glands, seminal vesicle, ejaculatory duct and/or testes affect female physiology, reproductive behavior, and longevity [16, 17]. The shape, size, and weight of the spermatophore made from the seminal fluids are greatly influenced by the structures of both male and female reproductive organs.

2. Function and structure of testes

Testes function is to produce sperm and in turn intake or excrete the various substances including nutrients and hormones for spermatogenesis. In species lacking seminal vesicles, mature sperm are stored in the testis near the vas deferens. The shape of the testis is circular, oval or elongated, perhaps corresponding to sperm length, probably because sperm develop greatly during spermiogenesis (from the spermatid to the sperm) [5]. The testes are usually paired and the numbers vary widely from species to species. The color of the testes also varies greatly depending on the species; for example, many lepidopteran insects have white, yellow, red, or purple testes. Although the pigments that are no longer needed in the body may be deposited in their testes, their function and evolutionary significance have not been clarified.

In *P. c-aureum*, the testis is completely covered with a yellow membrane from the last instar larval stage to the early adult stage while spermiogenesis is positively occurring [18]. Although the composition and function of this membrane are unknown, it is possible that it actively protects the testis and sperm cells from ultraviolet rays and/or various substances in the hemolymph, that are unsuitable or toxic to spermatogenesis. Alternatively, this membrane may positively improve the nutritional and/or humoral conditions for spermatogenesis. Interestingly, the time when this membrane begins to degenerate coincides with the time when adult development of wings is almost complete, the scale of spermatogenesis and testis size begin to shrink, and sperm begin to migrate from the testes to the vas deferens [18–20]. These synchronous events are strongly suspected to be associated with hormones in the hemolymph, such as ecdysteroid. Interestingly, it is shown in *Calpodes ethlius* Stoll (Lepidoptera, Hesperiidae) that the testis is surrounded by the yellow pigment [21].

Testis development including spermatogenesis is affected by developmental stages, temperatures, nutritional conditions, and hormones [22, 23]. In the yellow dung fly *Scathophaga stercoraria* (L.), the testes atrophy after mating [24]. Similarly, in *Drosophila melanogaster* Meigen, there was a significant reduction in the testes size after five successive matings [25]. In general, testis development is closely related to spermatogenesis [26], and testis size usually increases as spermatogenesis becomes more active [27]. However, the relationship between testis size and spermatogenesis depends on the species and developmental stage.

3. Sperm polyphenism

Sperm polymorphism is apparent in the various insect orders [1, 2]. Generally, one sperm is long whereas the other is short; giant sperm have been observed in some *Drosophila* and beetles. Giant sperm fertilize the eggs and short sperm do not participate in fertilization [28, 29], although contradictory findings raise the possibility that both morphs of sperm can fertilize [30]. In lepidopteran insects, there are fertile nucleated eupyrene sperm and infertile non-nucleated apyrene sperm [31–35]. Eupyrene sperm fertilize the eggs, whereas apyrene sperm cannot fertilize the eggs because they lose their nuclei during meiosis [36]. In general, apyrene sperm are produced more or transferred to females more than are eupyrene sperm [37–39], although spermatogenesis of eupyrene and apyrene sperm is not markedly different in the diamondback moth *Plutella xylostella* [40]. This seems to be related to the fact that apyrene sperm are shorter than eupyrene sperm and the former is generated later in development.

4. Storage and migration of sperm

The production and storage of sperm can be a lifelong event for males. There are four types of sperm migration within the male reproductive organs. First, sperm formed in the testis move to the vas deferens or the seminal vesicle. In lepidopteran insects, sperm migration occurs from the testis to the duplex via the vas deferens, with a circadian rhythm. Second, sperm move from the seminal vesicle through the ejaculatory duct to the female reproductive tract during mating. Third, there is a process called sperm reflux. In *P. c-aureum*, the accessory glands open into the duplex, but not into the ejaculatory duct, thus when the accessory gland material passes the duplex, it ejaculates all sperm present. This prevents sperm from being preserved in the next copulation. Thus this butterfly regurgitates excess sperm in the duplex into the vas deferens during mating. Finally, sperm migration sometimes occurs immediately after mating. Male sweetpotato weevils *Cylas formicarius* (F.) can mate several times a night, and sperm migrate from the testes to the seminal vesicles immediately after mating [41, 42]. This may be due to the fact that the seminal vesicles are adjacent to the testes, that spermatogenesis is active during the adult stage, and that sperm production is completed in the testes. Thus, the positional relationship between the testes and the seminal vesicles, the degree of sperm perfection in the testes, and the stage at which sperm are formed determine the male ejaculation pattern.

Sperm migration from the testis to the duplex via the vas deferens has been well studied in lepidopteran insects. Conversely, many dipteran insects do not have the seminal vesicles, which are sometimes called sperm reservoirs, however the distal end of the testis stores sperm. Their sperm are functionally complete. However, in lepidopteran insects, sperm are not mature in the testis and change morphologically when passing through the vas efferens from the testis to the vas deferens. It has been demonstrated that eupyrene sperm migration occurs in a circadian rhythm even in cases of in vitro culture of the testis- vas deferens-duplex complex [43]. It has also been reported that there is a circadian rhythm in the secretory activity of the upper vas deferens [44]. Although the reasons for sperm migration in lepidopteran insects being rhythmical are unclear, sperm migration is a time-consuming and energy-intensive process and thus it is reasonable to expect to be time managed.

Sperm formed in the testes are stored in the seminal vesicles until mating. At the time of mating, sperm are ejaculated into the female along with spermatophore. Interestingly, when *Heliothis virescens* (F.) are irradiated at the early adult stage, the sperm are not incorporated into the spermatophore because the sperm in the duplex has not moved to the simplex where the spermatophore is formed, resulting in no transfer of sperm [45]. Some studies have revealed that the so-called 'mating failure' often occurs in insects [46–48]. That is, they may mate but not pass spermatophore or sperm to females.

In *Drosophila melanogaster* Meigen, proteins within the seminal fluid of the male accessory gland are required for efficient accumulation of sperm in the female's sperm storage organs, and morphological changes in the shape and position and tissues within the female reproductive tract may be needed for successful sperm storage [49]. In species with a long life span, for example, the wood-feeding cockroach, *Cryptocercus punctutatus*, sperm are viable in the spermatheca for at least three years [50]. In these cases, female condition as well as the longevity of sperm is important for sustaining their survival.

Although the function of the seminal vesicles is to store and protect sperm, it is known that organs other than the seminal vesicles also store sperm. In lepidopteran insects, several species have a dilated part, also called a secondary seminal vesicle, between the lower vas deferens and the duplex [46, 51–53]. Even if all the sperm in the duplex are ejaculated during mating, these stored sperm can be replenished immediately from the secondary seminal vesicle. In addition, even species with a large swelling in the middle vas deferens may store sperm there to some extent temporarily [54], and may transfer sperm to the duplex after mating to prepare for subsequent matings. In species with these functions, sperm reflux would not be present.

5. Insemination

In insects, a spermatophore is passed into the female reproductive organs during mating directly or indirectly. The system of insemination varies greatly from species to species, the main cause of which is species diversity and complexity of the female reproductive organs. Probably, the female complex of reproductive organs have co-evolved with the male complex of reproductive organs.

The timing of transfer of sperm to females during mating varies greatly from species to species. Copula duration also varies widely, from minutes in parasitoid wasp and mosquitoes [55] to days in stick insects [17], but in the latter the ejaculation duration could be shorter than the copula one. The post-mating guard guarantees female oviposition behavior and ovarian development. It has been suggested that sperm progression to the spermatheca is supported by the activity of the muscles and nerves of the male and female reproductive organs in addition to sperm motility [56].

Fertilization efficiency also affects the number of ejaculations. As is well known, ants and wasps can efficiently fertilize many eggs with a small amount of sperm. For example, in a parasitoid wasp *Anisopteromalus calandrae* (Howard) (Hymenoptera: Pteromalidae), males have small (several hundred) numbers of sperm in the seminal vesicle and the fertilizing efficiency of stored sperm in the female genital organs is extremely high [57], although in the majority of insects the ratio of fertilized eggs/stored sperm is low due to polyspermy at fertilization. Orthopteran insects can ejaculate a bit at a time, probably because their seminal vesicles and accessory glands have an elongated gland-like structure and open separately into the ejaculatory duct. Therefore, in the desert locust, spermatophore and sperm can be gradually transferred to females gradually during mating from the proximal part of the gland near the ejaculatory opening. Because sperm migration starts to occur actively several days after adult emergence, it seems likely that new sperm are constantly stored near the entrance of the seminal vesicles and ejaculated into the females (Hiroyoshi, unpublished). This might be related to sperm aging and sperm competition. In the migratory grasshopper *Melanoplus sanguinipes* (F.), several spermatophores are passed to females in a single mating [58]. If the accessory glands and seminal vesicles were not elongated, it would be difficult to transfer spermatophore and sperm a bit at a time.

6. Positional relationship

The arrangement of male internal reproductive organs of various species is listed in **Table 1**. **Figure 1** shows the positional relationship between reproductive organs. Alphabetical order does not represent the phylogenetic relationship between species. Most probably other formats of organ positioning are yet to be discovered. Some insects lack the accessory glands. Of particular importance is the positional relationship between the testes, vas deferens, seminal vesicles, accessory glands, and ejaculatory duct, because the placement of these organs is important for ejaculation.

In type A, ejaculates reach the seminal vesicle or duplex form the testis via the vas deferens, to which the accessory glands are open. Thus far, all lepidopteran insects studied represent type A. As the accessory gland material passes through the duplex during mating, it flushes out all sperm and semen in the duplex. This relies on constant daily sperm replenishment form the testis to support multiple matings. Type B is similar to type A, but the seminal vesicles differ in that they connect to the ejaculatory duct along with the accessory glands. Type B is common in the homopteran, coleopteran and dipteran insects. In these insects, sperm production is typically completed in the testes, and in contrast to the Lepidoptera, the sperm do not require the vas deferens to mature. Type C lacks the seminal vesicles, but is basically the same as type B and more common in dipteran insects. Similarly, mature sperm can immediately be ejaculated from the testes during mating. Type C is an effective arrangement for multiple matings by adult males. Type D is found in some weevil species, where both the seminal vesicles and accessory glands are connected

Order	Family	Scientific name	Type	References
Thysanura	Lepismatidae	*Ctenolepisma campbelli*	F	[59]
Ephemeroptera	Leptophlebiidae	*Miroculis amazonicus*	H	[60]
Plecoptera	Taeniopterygidae	*Obipteryx sp.*	B	[61]
	Perlidae	*Perlesta placida*	G	[62]
Dermaptera	Labiduridae	*Labidura riparia*	H	[63]
	Anisolabididae	*Euborellia brunneri*	H	[64]
Isoptera	Termitidae	*Silvestritermes euamignathus*	H	[27]
Blattodea	Blattidae	*Leucophaea maderae*	E	[65]
Mantodea	Liturgusidae	*Ciulfina klassi*	E	[66]
Orthoptera	Caelifera	*Orphulella punctata*	E	[67]
	Tetrigidae	*Tetrixarenosa angusta*	E	[68]
	Pyrgomorphidae	*Poekilocerus pictus*	E	[69]
Psocoptera	Psyllipsocidae	*Dorypteryx domestica*	H	[70]
	Psoqvillidae	*Psoquilla marginepunctata*	H	[71]
Homoptera	Delphacidae	*Peregrinus maidis*	B	[72]
	Cicadellidae	*Graminella nigrifrons*	B	[73]
		Bothrogonia ferruginea	B	[74]
	Cicadoidea	*Tamasa tristigma*	B	[75]
Hemiptera	Reduviidae	*Rhodnius prolixus*	B	[76]
	Pentatomidae Aphididae	*Perillus bioculatus* *Acyrthosiphon pisum*	B C	[77] [78]
Hymenoptera	Colletidae	*Colletes rufipes*	F	[79]
	Andrenidae	*Oxaea flavencens*	F	[79]
	Megachilidae	*Anthidium manicatum*	B	[79]
	Apidae	*Centris violacea*	F	[79]
	Eulophidae	*Dahlbominus fuscipennis*	B	[80]
	Formicidae	*Solenopsis invicta*	F	[81]
Neuroptera	Chrysopidae	*Chrysopa oculata*	B	[82]
	Ithonidae	*Polystoechotes punctatus*	B	[83]
	Nemopteridae	*Palmipenna sp.*	B	[84]
Megaloptera	Corydalidae	*Parachauliodes continentalis*	B	[85]
Coleoptera	Rhysodidae	*Yamatosa nipponensis*	B	[86]
		Rhysodes comes	B	[87]
	Carabidae	*Campalita chinense*	C	[88]
		Damaster fruhstorferi	C	[88]
		Leistus prolongatus	C	[88]
	Tenebrionidae	*Parastizopus armaticeps*	B	[89]
	Trogossitidae	*Bolbocerosoma farctum*	B	[90]
		Meracantha contracta	B	[90]
	Chrysomelidae	*Galerucella birmanica*	B	[91]
		Leptinotarsa docemlineata	B	[92]

Order	Family	Scientific name	Type	References
	Elateridae	*Melanotus communis*	B	[90]
	Scarabaeidae	*Popillia japonica*	B	[90]
	Ciidae	*Hadraule blaisdelli*	B	[93]
	Curculionidae	*Anthonomus grandis*	B	[94]
		Dendroctonus armandi	B	[95]
	Bruchidae	*Bruchidius atrolineatus*	B	[96]
	Brentidae	*Cylas formicarius elegantulus*	D	[97]
Siphonaptera	Pulicidae	*Spilopsyllus cuniculi*	C	[98]
Diptera	Culicidae	*Anopheles gambiae*	B	[99]
	Psychodidae	*Culex pippiens* *Lutzomyia longipalpis*	B H	[100] [101]
	Tephritidae	*Anastrepha ludens*	C	[102]
		Strumeta tryoni	C	[103]
	Tachinidae	*Exorista sorbillans*	B	[104]
	Glossindae	*Glossina morsitans morsitans*	C	[105]
	Drosophilidae	*Drosophila melanogaster*	B	[106]
Lepidoptera	Crytophasidae	*Opisina arenosella*	A	[107]
	Olethreutidae	*Laspeyresia caryana*	A	[51]
	Pyralidae	*Ostrinia nubilalis*	A	[108]
	Tortricidae	*Choristoneura fumiferana*	A	[53]
	Noctuidae	*Trichoplusia ni*	A	[109]
		Pseudaletia unipuncta	A	[110]
		Agrotis ipsilon	A	[111]
		Heliothis aremigera	A	[112]
		Heliothis zea	A	[113]
		Spodoptera litura	A	[114]
	Geometridae	*Boarmia selenaria*	A	[115]
	Bombycidae	*Bombyx mori*	A	[116]
	Saturnidae	*Antheraea pernyi*	A	[117]

Table 1.
Classification of the male internal reproductive organs in various insects.

to the ejaculatory duct via the vas deferens. Type E is found in cockroaches and orthopteran insects, and through the testes the vas deferens leads to the tubular accessory glands and tubular seminal vesicles, both of which lead to the ejaculatory duct. In Type F, the testes open into the vas deferens or seminal vesicles and then the ejaculatory duct. Some hymenopteran insects belong to type F. Types G and H represent those insects that lack the accessory glands.

7. Conclusion

The position, as well as the morphology, structure and function of the male internal reproductive organs, is presumably adaptive, probably for successful

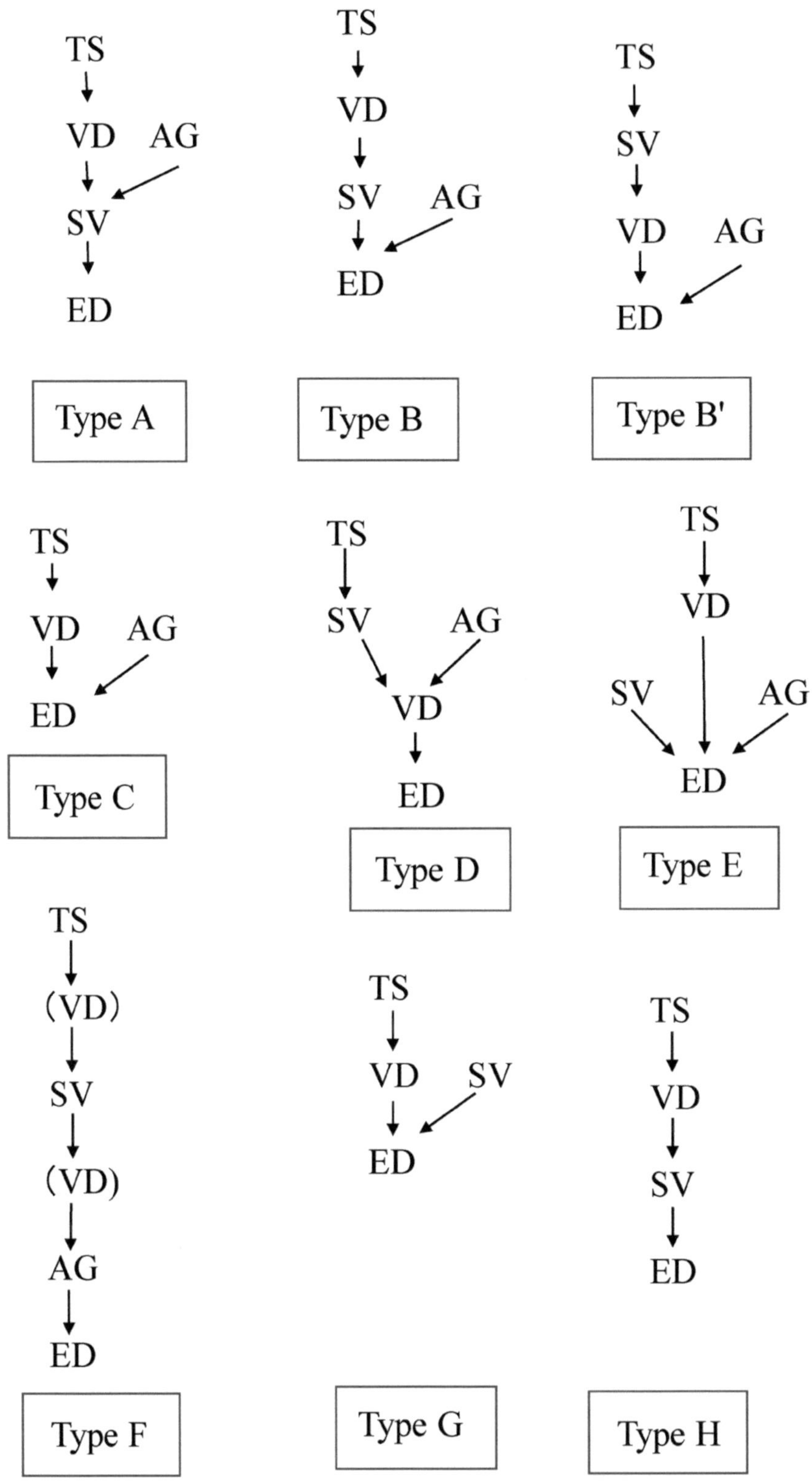

Figure 1.
Simple illustration of each type of configuration of male internal reproductive organs. TS, VD, SV, AG, and ED indicate testis, vas deferens, seminal vesicle, accessory gland, and ejaculatory duct, respectively. Arrow indicates the direction of ejaculates (semen or sperm).

reproduction. The complex folds of these organs, which are surrounded by the tracheae and fat bodies, suggests that they require large amounts of oxygen and nutrients. However, until now, research on the arrangement of the reproductive organs has not been prioritized. We hope that this chapter will serve as an opportunity and foundation for studying not only the morphology, structure and function of reproductive organs, but also their positional relationships.

Acknowledgements

This work is supported by USDA-ARS Research Project# 6066-22000-091-00D - Ecologically Sustainable Approaches to Insect Resistance Management in Bt Cotton.

Author details

Satoshi Hiroyoshi[1]* and Gadi V.P. Reddy[2]

1 Kawagoe, Saitama, Japan

2 USDA-ARS-Southern Insect Pest Management Research Unit, Stoneville, MS, USA

*Address all correspondence to: satoshi_hiroyoshi@yahoo.co.jp

IntechOpen

© 2021 The Author(s). Licensee IntechOpen. This chapter is distributed under the terms of the Creative Commons Attribution License (http://creativecommons.org/licenses/by/3.0), which permits unrestricted use, distribution, and reproduction in any medium, provided the original work is properly cited.

References

[1] Dallai R: Overview on spermato genesis and sperm structure of Hexapoda. Arthropod Structure & Development. 2014;43:257-290.

[2] Dallai R, Gottardo M, Beutel RG: Structure and evolution of insect sperm: New interpretations in the age of phylogenomics. 2016;61:1-23.

[3] Samani P: Digest: Evolution of sperm size and number in external fertilizers. Evolution. 2017;72:211-212.

[4] Hatsumi M, Wakahama KI: The sperm length and the testis length in *Drosophila nasuta* subgroup. Japanese Journal of Genetics. 1986;61: 241-244.

[5] Joly D. Bressac C: Sperm length in Drosophilidae (Diptera): Estimation by testis and receptacle lengths. International Journal of Insect Morphology & Embryology. 1994;23:85-92.

[6] Pitnick S: Investment in testes and the cost of making long sperm in *Drosophila*. The American Naturalist. 1996;148:57-80.

[7] Gage MJG: Associations between body size, mating pattern, testis size and sperm lengths across butterflies. Proceedings of the Royal Society of London B. 1994;258:247-254.

[8] Joly D, Bazin C, Zeng LW, Singh RS: Genetic basis of sperm and testis length differences and epistatic effect on hybrid inviability and sperm motility between *Drosophila simulans* and *D. sechellia*. Heredity. 1995;78:354-362.

[9] Immler S, Pitnick S, Parker GA, Durrant KL, Lüpold S, Calhim S, Birkhead TR: Resolving variation in the reproductive tradeoff between sperm size and number. Proceedings of the National Academic Sciences of the United States of America. 2011;108:5325-5330.

[10] Mojica JM, Bruck DL: Sperm bundle coiling: Transporting long sperm bundles in *Drosophila dunni dunni*. Journal of Insect Physiology. 1996;42:303-307.

[11] Pizzari T: Evolution: sperm ejection near and far. Current Biology. 2004;14:R511-R513.

[12] Sepil I, Hopkins BR, Dean R, Bath E, Friedman S, Swanson B, Ostridge, HJ, Harper L, Buehner NA, Wolfner MF, Konietzny R, Thézénas ML, Sandham E, Charles PD, Fischer B, Steinhauer J, Kessler BM, Wigby S: Male reproductive aging arises via multifaceted mating-dependent sperm and seminal proteome declines, but is postponable in *Drosophila*. Proceedings of the National Academic Sciences of the United States of America. 2020;117:17094-17103.

[13] Strobl V, Sstraub L, Bruckner S, Albrecht M, Maitip J, Kolari E, Chantawannakul P, Williams GR, Neumann P: Not every sperm counts: Male fertility in solitary bees, *Osmia cornuta*. PLoS ONE. 2019;14:e0214597.

[14] Simmons LW, Lovegrove M: Socially cued seminal fluid gene expression mediates responses in ejaculate quality to sperm competition risk. Proceedings of the Royal Society B. 2017;284:1486.

[15] Hiroyoshi S: Regulation of sperm quantity transferring to females at mating in the adult male of *Polygonia c-aureum* (Lepidoptera: Nymphalidae). Applied Entomology and Zoology. 1995;30:111-119.

[16] Sirot LK, Lapointe SL, Shatters R, Bausher M: Transfer and fate of seminal fluid molecules in the beetle, *Diaprepes abbreviatus*: Implications for the reproductive biology of a pest species.

Journal of Insect Physiology. 2006;52:300-308.

[17] Chapman RF: SJ, Simpson SJ, Douglas AE, editors. The insects. Structure and function. 5th ed. UK: Cambridge University Press; 2013. 929p.

[18] Hiroyoshi S: Effects of aging, temperature and photoperiod on testis development of *Polygonia c-aureum* (Lepidoptera: Nymphalidae). Entomological Science. 2000;3:227-236.

[19] Hiroyoshi S: Effects of photoperiod and age on the initiation of sperm movement in male *Polygonia c-aureum* Linnaeus (Lepidoptera: Nymphalidae). Applied Entomology and Zoology. 1997;32:19-25.

[20] Hiroyoshi S, Reddy GVP, Mitsunaga T: Effects of photoperiod and aging on the adult spermatogenesis of *Polygonia c-aureum* (Lepidoptera: Nymphalidae), in relation to adult diapause. Journal of Comparative Physiology A. 2020;206:467-475.

[21] Lai-Fook J: Testicular development and spermatogenesis in *Calpodes ethlius* Stoll (Hesperiidae, Lepidoptera). Canadian Journal of Zoology. 1982;60:1161-1171.

[22] Dumser JB: The regulation of spermatogenesis in insects. Annual Review of Entomology. 1980;25: 341-369.

[23] Droney DC: The influence of the nutritional content of the adult male diet on testis mass, body condition and courtship vigour in a Hawaiian *Drosophila*. Functional Ecology. 1998;12:920-928.

[24] Ward PI, Simmons LW: Copula duration and testes size in the yellow dung fly, *Scathophaga stercoraria* (L.): the effects of diet, body size, and mating history. Behavioral Ecology and Sociobiology. 1991;29:77-85.

[25] Linklater JR, Wertheim B, Wigby S, Chapman T: Ejaculate depletion patterns evolve in response to experimental manipulation of sex ratio in *Drosophila melanogaster*. Evolution. 2007;61:2027-2034.

[26] Du Q, Wen L, Zheng SC, Bi HL, Huang YP, Feng QL, Liu L: Identification and functional characterization of *doublesex* gene in the testis of *Spodoptera litura*. Insect Science. 2019;26:1000-1010.

[27] Laranjo LT, Haifig I, Costa-Leonardo AM: Morphology of the male reproductive system during post-embryonic development of the termite *Silvestritermes euamignathus* (Isoptera: Termitidae). Zoologischer Anzeiger. 2018;272:20-28.

[28] Snook RR, Markow TA: Possible role of nonfertilizing sperm as a nutrient source for female *Drosophila pseudoobscura frolova* (Diptera: Drosophilidae). Pan-Pacific Entomologist. 1996;72:121-129.

[29] Snook, RR: Is the production of multiple sperm types adaptive? Evolution. 1997;51:797-808.

[30] Bressac C, Hauschteck-Jungen E: *Drosophila subobscura* females preferentially select long sperm for storage and use. Journal of Insect Physiology. 1996;42:323-328.

[31] Silbergried RE, Shepherd JG, Dickinson JL: Eunuchs: The role of apyrene sperm in Lepidoptera? The American Naturalist. 1984;123:255-265.

[32] Friedländer M, Seth RK, Reynolds SE: Eupyrene and apyrene sperm: Dichotomous spermatogenesis in Lepidoptera. Advanced Insect Physiology. 2005;32:206-308.

[33] Friedländer M, Miesel S: Spermatid anucleation during the normal atypical spermatogenesis of the warehouse moth

Ephestia cautella. Journal of Submicroscopic Cytology 1977;9:173-185.

[34] Mongue AJ, Hansen ME, Gu L, Sorenson CE, Walters JR: Nonfertilizing sperm in Lepidoptera show little evidence for recurrent positive selection. Molecular Ecology. 2019;28.

[35] Esfandi K, He XZ, Wang Q: Sperm allocation strategies in a sperm heteromorphic insect. Current Zoology. 2020;66:285-292.

[36] Chen S, Liu Y, Yang X, Liu Z, Luo X, Xu J, Huang Y: Dysfunction of dimorphic sperm impairs male fertility in the silkworm. Cell Discovery. 2020;6:60.

[37] Hiroyoshi S: Eupyrene and apyrene spermatogenesis in the Asian comma butterfly, *Polygonia c-aureum* (Lepidoptera: Nymphalidae). Entomological Science. 1999;2:297-305.

[38] TengZ-Q, Zhang Q-W: Determinants of male ejaculate investment in the cotton bollworm *Helicoverpa armigera*: mating history, female body size and male age. Physiological Entomology. 2009;34:338-344.

[39] Win AT, Ishikawa Y: Effects of diapause on post-diapause reproductive investment in the moth *Ostrinia scapulalis*. Entomologia Experimentalis et Applicata. 2015;157:346-353.

[40] Li X, Zhang K, Deng Y, He R, Zhang X, Zhong G, Hu Q, Weng Q: Effects of ^{60}Co-γirradiation on testis physiological aspects of *Plutella xylostella* (Linnaeus). Ecotoxicology and Environmental Safety. 2019;169: 937-943.

[41] Hiroyoshi S, Kohama T, Reddy GVP: Age-related sperm production, transfer, and storage in the sweet potato weevil, *Cylas formicarius* (Fabricius) (Coleoptera: Curculionidae). Journal of Insect Behavior. 2016;29:689-707.

[42] Hiroyoshi S, Reddy GVP, Kohama T: Sperm supply from the testes to the seminal vesicle over consecutive matings in the sweetpotato weevil, *Cylas formicarius* (FABRICIUS) (Coleoptera: Curculionidae). American Journal of Life Sciences. 2014;5:103-107.

[43] Giebultowicz JM, Riemann JG, Raina AK, Ridgway RL: Circadian system controlling release of sperm in the insect testes. Science. 1989;245:1098-1100.

[44] Bebas P, Maksimiuk E, Gvakharia B, Cymborowski B, Giebultowicz JM: Circadian rhythm of glycoprotein secretion in the vas deferens of the moth, *Spodoptera littoralis*. BMC Physiology. 2002;2:15

[45] Flint HM, Kressin EL: Transfer of sperm by irradiated *Heliothis virescens* (Lepidoptera: Noctuidae) and relationship to fecundity. The Canadian Entomologist. 1969;101:500-507.

[46] Drummond BA: Multiple mating and sperm competition in the Lepidoptera. In: Smith RL, editor. Sperm competition and the evolution of animal mating systems. London: Academic Press; 1984. P. 291-360.

[47] Rhainds M: Female mating failures in insects. Entomologia Experimentalis et Applicata.;2010;136:211-226.

[48] Balfour VL, Black D, Shuker, DM: Mating failure shapes the patterns of sperm precedence in an insect. Behavioral Ecology and Sociobiology. 2020;74:25.

[49] Adams EM, Wolfner MF: Seminal proteins but not sperm induce morphological changes in the *Drosophila melanogaster* female reproductive tract during sperm storage. Journal of Insect Physiology. 2007;53:319-331.

[50] Nalepa CA, Mullins DE: Repeated copulation in the wood-feeding cockroach *Cryptocercus punctulatus* does not influence number or development of offspring. Journal of Insect Behavior. 2011;24:44-54.

[51] Tedders Jr WL, Calcote VR: Male and female reproductive systems of *Laspeyresia caryana*, the hickory shuckworm moth (Lepidoptera: Olethreutidae). Annals of the Entomological Society of America. 1967;60:280-282.

[52] Tedders Jr WL, Osburn M: Morphology of the reproductive systems of *Gretchena bolliana*, the pecan bud moth. Annals of the Entomological Society of America. 1970;63:786-789.

[53] Outram I: Morphology and histology of the reproductive system of the male spruce budworm, *Choristoneura fumiferana*. The Canadian Entomologist. 1970;102:404-414.

[54] Ferro DN, Akre RD: Reproductive morphology and mechanics of mating of the codling moth, *Laspeyresia pomonella*. Annals of the Entomological Society of America. 1975;68:417-424.

[55] Boes KE, Ribeiro JMC, Wong A, Harrington LC, Wolfner MF, Sirot LK: Identification and characterization of seminal fluid proteins in the Asian tiger mosquito, *Aedes albopictus*. PLOS Neglected Tropical Diseases. 2014;8:e2946.

[56] Barcellos MS, Martins LCB, Cossolin JFS, Serrão JE, Delabie JHC, Lino-Neto J: Testes ant spermatozoa as characters for distinguishing two species of the genus *Neoponera* (Hymenoptera: Formicidae). Florida Entomologist. 2015;98:1254-1256.

[57] Bressac C, Khanh HDT, Chevrier C: Effects of age and repeated mating on male sperm supply and paternity in a parasitoid wasp. Entomologia Experimentalis et Applicata. 2009;130:207-213.

[58] Pickford R, Gillott C: Insemination in the migratory grasshopper, *Melanoplus sanguinipes* (Fabr.). Canadian Journal of Zoology. 1971;49:1583-1588.

[59] Barnhart Sr CS: The internal anatomy of the silverfish *Ctenolepisma campbelli* and *Lepisma saccharinum* (Thysanura: Lepismatidae). Annals of the Entomological Society of America. 1961;54:177-196.

[60] Brito P, Salles FF, Dolder H: Characteristics of the male reproductive system and spermatozoa of Leptophlebiidae (Ephemeroptera). Neotropical Entomology. 2011;40:103-107.

[61] Endoh S, Niwa N, Matsuzaki M: Comparative anatomy of the male reproductive organs and a brief description on the spermatozoon structure of *Kamimuria quadrata* (Plecoptera). Fukushima University. Science Report. 1995;56:13-23.

[62] Stewart KW, Atmar GL, Solon BM: Reproductive morphology and mating behavior of *Perlesta placida* (Plecoptera: Perlidae). Annals of the Entomological Society of America. 1969;62:1433-1438.

[63] Kamimura Y: Right-handed penises of the earwig *Labidura riparia* (Insecta, Dermaptera, Labiduridae): Evolutionary relationships between structural and behavioral asymmetries. Journal of Morphology. 2006;267:1381-1389.

[64] van Lieshout E: Male genital length and mating status differentially affect mating behaviour in an earwig. Behavioral Ecology and Sociobiology. 2011;65:149-156.

[65] van Wyk LE: The morphology and histology of the genital organs of Leucophaea maderae (Fabr.) (Blattidae,

Orthoptera). Journal of the Entomological Society of South Africa. 1952;15:1-62.

[66] Winnick CG, Holwell GI, Herberstein ME: Internal reproductive anatomy of the praying mantid *Ciulfina klassi* (Mantodea: Liturgusidae). Arthropod Structure & Development. 2009;38:60-69.

[67] Silva DSM, Cossolin JFS, Pereira MR, Lino-Neto J, Sperber CF, Serrão JE: Male reproductive tract and spermatozoa ultrastructure in the grasshopper *Orphulella punctata* (De Geer, 1773) (Insecta, Orthoptera, Caelifera). Microscopy Research Technique. 2017:1-6.

[68] Widdows RE, Wick JR: Morphology of the reproductive system of *Tetrix arenosa angusta* (Hancock) (Orthoptera, Tetrigidae). Proceedings of the Iowa Academy of Science. 1959;66:484-503.

[69] Wagan MS, Pitafi KD: The anatomy and histology of male reproductive organs of *Poekilocerus pictus* (Fabricius) (Pyrgomorphidae: Acridoidea: Orthoptera). Pakistan Journal of Zoology. 1990;22:117-121.

[70] Golub NV, Kučerová Z: Karyotype and reproductive organs of male *Dorypteryx domestica* (Smithers, 1958) (Psocoptera: Trogiomorpa: Psyllipsocidae). Folia biologica (Kraków). 2008;56:21-23.

[71] Kai WS, Thornton IWB: The internal morphology of the reproductive systems of some psocid species. Proceedings of the Royal Society of London (A). 1968;43:1-12.

[72] Tsai JH, Perrier JL: Morphology of the digestive and reproductive systems of *Peregrinus maidis* (Homoptera: Delphacidae). Florida Entomologist. 1993;76:428-436.

[73] Tsai JH, Perrier JL: Morphology of the digestive and reproductive systems of *Dalbulus maidis* and *Graminella nigrifrons* (Homoptera: Cicadellidae). Florida Entomologist. 1996;79: 563-578.

[74] Hayashi F, Kamimura Y: The potential for incorporation of male derived proteins into developing eggs in the leafhopper *Bothrogonia ferruginea*. Journal of Insect Physiology. 2002;48:153-159.

[75] Moulds MS: An appraisal of the higher classification of cicadas (Hemiptera: Cicadoidea) with special reference to the Australian fauna. Records of the Australian Museum. 2005;57:375-446.

[76] Khalifa A: Spermatophore production and egg-laying behaviour in *Rhodnius prolixus* Stal. (Hemiptera; Reduviidae). Parasitology. 1950;40:283-289.

[77] Adams TS: Morphology of the internal reproductive system of the male and female two-spotted stink bug, *Perillus bioculatus* (F.) (Heteroptera: Pentatomidae) and the transfer of products during mating. Invertebrate Reproduction and Development. 2001;39:45-53.

[78] Wieczorek K, Kanturski M, Sempruch C, Świątek P: The reproductive system of the male and oviparous female of a model organism- the pea aphid, Acythosiphon pisum (Hemiptera, Aphididae). PeerJ. 2019;7:e7573.

[79] Ferreira A, Avdalla FC, Kerr WE, da Cruz-Landim C: Comparative anatomy of the male reproductive internal organs of 51 species of bees. Neotropical Entomology. 2004;33:569-576.

[80] Araújo VA, Freitas FV, Moreira J, Neves CA, Lino-Neto J.: Morphology of male reproductive system of two solitary

bee species (Hymenoptera: Apidae). Neotropical Entomology. 2010;39.

[81] Wilkes A: Sperm transfer and utilization by the arrhenotokous wasp *Dahlbominus fuscipennis* (Zett.) (Hymenoptera: Eulophidae). The Canadian Entomologist. 1965;97:647-657.

[82] Ball DE, Vinson SB: Anatomy and histology of the male reproductive system of the fire ant, *Solenopsis invicta* Buren (Hymenoptera: Formicidae). International Journal of Insect Morphology & Embryology. 1984;13:283-294.

[83] Hwang JC, Bickley WE: The reproductive system of *Chrysopa oculata* (Neuroptera: Chrysopidae). Annals of the Entomological Society of America. 1961;54:422-427.

[84] de Jong GD: Observations on the biology of *Polystoechotes punctatus* (Fabricius) (Neuroptera: Ithonidae): Adult trophic status, description of the male reproductive system, and associations with mites. Proceedings of the Entomological Society of Washington. 2011;113:291-298.

[85] Walker MH, Picker MD, Leon B: Eversible abdominal vesicles and some observations of the male reproductive system of the spoon wing lacewing *Palmipenna* (Neuropera: Nemopteridae). Journal of Morphology. 1994;219:47-58.

[86] Hayashi F: Insemination through an externally attached spermatophore: Bundled sperm and post-copulatory mate guarding by male fishflies (Megaloptera: Corydalidae). Journal of Inset Physiology. 1996;42:859-866.

[87] Yahiro K: Comparative morphology of the alimentary canal and reproductive organs of the terrestrial Caraboidea (Coleoptera: Adephaga) Part 1. Japanese Journal of Entomology. 1996;64:536-550.

[88] Yahiro K: Comparative morphology of the alimentary canal and reproductive organs of the terrestrial Caraboidea (Coleoptera: Adephaga) Part 2.Entomological Science. 1998;1:47-53.

[89] Brits JA: The anatomy, histology and physiology of the internal adult male reproductive system of *Parastizopus armaticeps* Péringuey (Coleoptera: Tenebrionidae). Journal of the Entomological Society of south Africa. 1982;45:239-260.

[90] Williams JL: The anatomy of the internal genitalia of some Coleoptera. Proceedings of the Entomological Society of Washington. 1945;47:73-91.

[91] Verma KK: Functional and developmental anatomy of the reproductive organs in the male of *Galerucella birmanica* Jac (Coleoptera, Phytophaga, Chrysomelidae). Annales des Sciences Naturelles, Zoologie, Paris. 1969;12:139-234.

[92] Smid HM: Transfer of a male accessory gland peptide to the female during mating in *Leptinotarsa decemlineata.* Invertebrate Reproduction and Development. 1998;34:47-53.

[93] Klopfenstein WG, Graves RC: Morphology of the digestive and reproductive systems of adult *Hadraule blaisdelli* (Casey) (Coleoptera: Ciidae). The Coleoptersist Bulletin. 1992;46:344-356.

[94] Burke HR: Morphology of the reproductive systems of the cotton boll weevil (Coleoptera, Curculionidae). Annals of the Entomological Society of America. 1959;52:287-294.

[95] Wu YF, Wei LS, Torres MA, Zhang X, Wu S-P, Chen H: Morphology of the male reproductive system and

spermiogenesis of *Dendroctonus armandi* (Coleoptera: Curculionidae: Scolytinae). Journal of Insect Science. 2017;17:1-9.

[96] Glitho IA, Huignard J: A histological and ultrastructural comparison of the male accessory reproductive glands of diapausing and non-diapausing adults in *Bruchidius atrolineatus* (Pic) (Coleoptera: Bruchidae). International Journal of Insect Morphology & Embryology. 1990;19:195-209.

[97] Calder AA: Gross morphology of the soft parts of the male and female reproductive systems of Curculionoidea (Coleoptera). Journal of Natural History. 1990;24:453-505.

[98] Mead-Briggs AR: The structure of the reproductive organs of the European rabbit-flea, *Spilopsyllus cuniculi* (Dale) (Siphonaptera). Proceedings of the Royal Society of London (A). 1962;37:79-88.

[99] Huho BJ, Ng'habi KR, Killeen GF, Nkwengulila G, Knols BGJ. Ferguson HM: A reliable morphological method to assess the age of male *Anopheles gambiae*. Malaria Journal. 2006;5:62.

[100] Meuti ME, Short SM: Physiological and environmental factors affecting the composition of the ejaculate in mosquitoes and other insects. Insects 2019;10:74.

[101] Spiegel CN, Bretas JAC, Peixoto AA, Vigoder FM, Bruno RV, Soares MJ: Fine structure of the male reproductive system and reproductive behavior of *Lutzomyia longipalpis* sandflies (Diptera: Psychodidae: Phlebotominae). PLoS ONE. 2013:8:e74898.

[102] Valdez JM: Ultrastructure of the testis of the Mexican fruit fly (Diptera: Tephritidae). Annals of the Entomological Society of America. 2001;94:251-256.

[103] Drew RAI: Morphology of the reproductive system of *Strumeta tryoni* (Froggatt) (Diptera: Trypetidae) with a method of distinguishing sexually mature adult males. Journal of the Australian Entomological Society; 1969;8:21-32.

[104] Veeranna G, Prasad NR: Reproductive biology of uzi fly *Exorista sorbillans* (Diptera: Tachinidae). In: Recent advances in uzi fly research; 16-17 January 1992; Bangalore, India; 1993.p13-22

[105] Odhiambo TR, Kokwaro ED, Sequeira LM: Histochemical and ultrastructural studies of the male accessory reproductive glands and spermatophore of the tsetse, *Glossina morsitans morsitans* Westwood. Insect Science and its Application. 1983;4: 227-236.

[106] Ram KR, Wolfner MF: Seminal influences: *Drosophila* Acps and the molecular interplay between males and females during reproduction. Integrative and Comparative Biology. 47;2007:427-445.

[107] Santhosh-Babu PB: Development and differentiation of male reproductive organs in *Opisina arenosella* Walker. Entomon. 1995;20:59-66.

[108] Jones JA, Guthrie WD, Brindley TA: Postembryonic development of the reproductive system of male European corn bores, *Ostrinia nubilalis* (Lepidoptera: Pyralidae). Annals of the Entomological Society of America. 1984;77:155-164.

[109] Holt GG, North DT: Effects of gamma irradiation on the mechanisms of sperm transfer in *Trichoplusia ni*. Journal of Insect Physiology. 1970;16:2211-2222.

[110] Callahan PS, Chapin JB: Morphology of the reproductive systems and mating in two representative

members of the family Noctuidae, *Pseudaletia unipuncta* and *Peridroma margaritosa*, with comparison to *Heliothis zea*. Annals of the Entomological Society of America. 1960;53:763-782.

[111] Gemeno C, Anton S, Zhu JW, Haynes KF: Morphology of the reproductive system and antennal lobes of gynandromorphic and normal black cutworm moths, *Agrotis ipsilon* (Hufnagel) (Lepidoptera: Noctuidae). International Journal of Insect Morphology & Embryology. 1998;27:185-191.

[112] Hoque MR: Comparative morphology of the reproductive systems of *Heliothis armigera* (Hubner) and *Heliothis punctigera* Wallengren (Lepidoptera: Noctuidae). Bangladesh Journal of Zoology. 1992;20:17-26.

[113] Callahan PS: Serial morphology as a technique for determination of reproductive patterns in the corn earworm, *Heliothis zea* (Boddie). Annals of the Entomological Society of America. 1958;51:413-428.

[114] Etman AAM, Hooper GHS: Developmental and reproductive biology of *Spodoptera litura* (F.) (Lepidoptera: Noctuidae). Journal of the Australian Entomological Society. 1979;18:363-372.

[115] Scheepens MHM, Wysoki M: Reproductive organs of the giant looper, *Boarmia selenaria* Schiffermüller (Lepidoptera: Geometridae). International Journal of Insect Morphology & Embryology. 1986;15:73-81.

[116] Omura S: Studies on the reproductive system of the male of *Bombyx mori*, IV.Post-testicular organs and post-testicular behaviour of the spermatozoa. Journal of Faculty of Agriculture of Hokkaido Imperial University Sapporo. 1938;40: 129-170.

[117] Shepherd JG: Activation of saturniid moth sperm by a secretion of the male reproductive tract. Journal of Insect Physiology. 1974;20:2107-2122.

Section 2

Reproductive Toxicology

Chapter 5

Reproductive Toxicology: An Update

Makhadumsab Toragall, Shridhar C. Ghagane, Rajendra B. Nerli and Murigendra B. Hiremath

Abstract

Human reproduction and development is a succession of symbiotic events. Nearly, at every point of this phenomenon found to be the principle target of one or more reproductive toxicants. Chemical agents, physical factors, as well as biological intruders can pose antagonistic effects on reproductive potential of an organism. The pathways are different *viz.*, either damaging embryo and sometimes fetus or inducing mutation in a parent's germ cell. The outcomes are declined fertility to impulsive abortion, functional discrepancies, developmental retardation, structural anomalies, etc. It is a now essential to establishing proper databases for reproductive and developmental toxicity chemicals, physical and biological factors including appropriate awareness among the society. Although many *in vitro* and *in vivo* toxicology studies are in pipeline which are independent studies but combination with other hazardous studies could give us an accurate numbers.

Keywords: reproduction, toxicity, fertility, infertility, mutagens

1. Introduction

Reproduction (procreation; conception) is one of the most essential requirements of all organisms where producing a transcript, helps in the survival and perpetuation of the species [1]. Parenthood is one of the most comprehensively preferred priorities of mankind and which happens at ease, when couples are vigorous and normal. According to earlier archaic studies of anthropologists and evolutionary biologists, *Homo sapiens* displayed a better cognitive development about 2,00,000 years ago and hence, had a "reproductive consciousness" [2]. Ancient mythologies and civilizations respected and worshipped fertility Gods and Goddesses like, Egyptian goddess *Maat* [3], Mesopotamian's *Erua* [4], Babylonian's *Ishtar*, Persian's *Anaitis*, Greek's *Actemia* [2], and also they had a deep desire for conception, and a strong perception of fertility, which can be correlated with human sustainability and existence.

The reproductive cycle of a mammalian individual involves of a sequence of several phases and unified events. According to ICH guideline [5]:

A. Pre-mating to birth (mature male and female conceptive ability, growth and maturation of gametes, reproductive nature, and conception).

B. Conception to implantation (mature female conceptive ability, pre-implantation development, cleavage, morula, blastula, and implantation).

C. Implantation and organogenesis (mature female conceptive ability, development of embryo, and foremost organ development).

D. Fetal development (until the end of gestation, mature female conceptive ability, fetal development and growth, and growth and development of organs).

E. Birth and pre-weaning development (mature female conceptive ability, parturition, lactation, neonate adaptation to extrauterine life, pre-weaning development, and growth).

F. Post-weaning development up to sexual maturity (growth, adulthood, adaptation to independent life, and achievement to full sexual function).

According to World Health Organization (WHO) info, at present globally 50–80 million people are facing infertility [6]. Significant studies have reported that female infertility occurs 50%, infertility because of male factors is 20–30%, and rest is shared by both genders [7]. These findings are considerably broader than previously reported.

2. Reproductive toxicity

Since a decade, human reproductive disruption by various factors including xenobiotics such as drugs, occupational, and environmental exposures leading to reproductive toxicity which is has become a growing concern. Reproductive toxicity defined as: "the antagonistic effects of a substance on any characteristics of the male or female sexual reproductive cycle, together with an impairment of reproductive function, and the induction of adverse effects in the embryo, such as growth retardation, malformations, and death which would interfere with the production and development of normal offspring that could be reared to sexual maturity, capable in turn of reproducing the species" [8].

The first essential introduction in reproductive toxicology was given by Wilson and Warkany in 1965 [9]. The first test guideline was published by the Food and Drug Administration (FDA) in 1966 [10], followed by the Committee on Safety of Medicines [11], Ministry of Health and Welfare (MHW) of Japan [12], and rest of the other nations. It was provisionally terminated by International Federation of Teratology Societies (IFTS), pharmaceutical industry and the health authorities of EEC, Japan, and USA with the aid of ICH Harmonized Tripartite Guideline "Detection of Toxicity to Reproduction for Medicinal Products" in June 1993 [13, 14].

Reproductive toxicity is categorized as follows:

2.1 Reproductive toxicity

Reproductive toxicity has been defined as "any effect of chemicals that would interfere with reproductive ability or potential," with consequent effects on lactation and the development of the progenies [15]. It includes, variations in the reproductive system of men and women, adverse impacts on the beginning of adolescence, normal reproductive cycle, production and transport of healthy gametes, sexual activities, fertility, parturition, early conceptive senescence, and alterations in any other activities which are reliant on the integrity of the reproductive systems [16]. Reproductive toxicity effects could be via lactation too but such classes are treated separately [17]. This is because it is desirable to be able to classify chemicals specifically for adverse effect on lactation so that a specific hazard warning about this effect can be provided for lactating mothers.

Classes of reproductive toxicity include:

- Male fertility
- Female fertility
- Parturition
- Lactation.

2.2 Developmental toxicity

According to Globally Harmonized System the developmental toxicity is defined as, “adverse effects induced during pregnancy, or as a result of parental exposure,” which “can be manifested at any point in the life span of the organism” [15]. The exposure to specific exogenous substances prior to conception in either of the parent, exposure during gestation, or exposure during prenatal or postnatal development from birth to sexual maturation may result in developmental toxicity. Developmental toxicity has varied end points such as impulsive abortions, still-births, deformities, and early postnatal mortality, reduced birth weight leading to structural anomaly, altered growth, functionally deficit, and death of the developing organism [18].

Classes of developmental toxicity include:

- Mortality
- Dysmorphogenesis (structural abnormalities)
- Alterations to growth
- Functional impairment.

Due to the fact that, male and female reproductive anatomy and biologic mechanisms are differing, they have a speckled result for reproductive toxicants. It is therefore essential to recognize reproductive toxins and their mechanisms and sites of action and to learn about species (especially human) vulnerability to them. Reproductive toxicants or reprotoxicant are chemical, biohazardous (e.g., viruses), or physical (e.g., radiation), agents that can impair the reproductive capabilities in men and/or women. Developmental toxicants interfere with proper growth or health of the child acting at any point from conception to puberty. The chemical agents which elevate the occurrence mutations above natural level by damaging the genetic material of an individual are known as mutagens. Incidences of defective cells or cancerous cells found when these mutations are inherited. As the name suggests, embryotoxins are lethal to embryos, where they may exterminate, distort, impede the growth and development of embryo, and may cause postnatal problems. The compounds like, mercury, lead, other heavy metals, and organic compounds *viz.*, formamide are some of the well-known examples of embryotoxins. Additionally, agents which can interrupt or leads to deformity in the development of an embryo or fetus are called as teratogens, which have the potential to miscarriage or cause children with birth defects.

The fundamental biological mechanisms of reproductive toxicity are multifaceted and involve absorption, distribution, metabolism (toxification, and/or detoxification), excretion, and repair [19]. The mechanism of reproductive toxicants disturbs the flow of matter, energy, or information that are necessary for normal functioning of cells, organs, or organisms. Later, toxicant will distribute to the target organ (gonad, hypothalamus, pituitary, uterus, epididymis, liver, etc.) where it employs its antagonistic effect before it is metabolized. The action of reproductive toxicant is either direct by the virtue of structural similarity to an endogenous compound (*viz.*, hormone, vitamin, or nutrient) or because of chemical reactivity (i.e., alkylating agent, denaturant, and chelator) or indirect (requiring metabolism before exerting a toxic effect) on reproductive system [19].

Formerly debated male fertility decline is no longer controversial. Several substantiating reports have confirmed a fall in sperm counts and semen quality in men over the last several decades globally [20]. The causative factors are not only obesity, illicit substance use, smoking rates, and alcohol abuse, but also due to chronic reproductive toxin exposures of the modern age. The spermatozoa with abnormal genetic material, which supports irregular spermatogenesis, abortions, progenies with genetic defect and diseases, etc., are some of the adverse effect of causative factors on male fertility. The investigation of male reproductive function begins with measurement of testis size, semen analysis, accessory gland functionality, reproductive hormone estimation, impotence or reduced libido, and fertility [21].

Many studies have been reporting weakening of female reproductive capacity over the past half century, which could be because of cultural change (e.g., delayed childbearing and increased contraception in women), but environmental exposures to the fetus, mother, or father may also contribute [22]. The women's health practitioners, obstetricians, and gynecologists have been advised to increase communication with their patients about the potentially detrimental effects of reproductive toxicants on reproductive health [23]. Among women of US and Danish the conception rate have declined to 44% since 1960 [24] and hormone-related diseases such as disorders of pubertal development, polycystic ovary syndrome (PCOS), endometriosis, and uterine fibroids have become common.

Reproductive difficulties and developmental abnormalities constitute a significant medical problem and greatly contribute to human suffering. The following provides a summary of the Globally Harmonized System (GHS) system as it relates to classification of health hazards [15]. The GHS system defines developmental toxicity with reproductive toxicity, but later classifies them separately. While classification, the GHS system define reproductive toxicity as, antagonistic effects on sexual activity and fertility amongst adult men and women comprising its adverse effect on sexual deeds, parturition, pregnancy outcomes [15].

These are the following categories of reproductive and developmental toxicants:

2.2.1 Category I: satisfactory reports with human evidence

The toxicants which are found to be human reproductive and/or developmental hazards are considered as category-I agents. The research reports which assist the above hypothesis with satisfactory epidemiologic confirmations or studies involving humans along with solid subsidiary animal study for at least one adverse reproductive effect. Though, the research data with human study is limited to support this classification, at present there are agents which fall under this category. The list includes biological, physical along with chemical hazards which pose potential reproductive effects, are the resultant of studies included humans and animals (**Table 1**).

Chemical	Reported adverse effects
Aniline	Female subfertility, natural abortion, growth impedance, and developmental disorders
Bulsulfan, methotrexate, cyclophosphamide	Male and female infertility, natural abortion, genetic defects, and growth retardation
Carbon disulfide	Lower male sex libido, infertility, abortion, abnormal growth, menstrual disorders, and breast milk contamination
Carbon monoxide	Female infertility, spontaneous abortion, growth retardation, and functional deficit
Dibromochloropropane (DBCP)	Infertility in men, genetic defects, and altered sex ratios
Dinitrotoluene (DNT)	Spontaneous abortion, male infertility, growth retardation, and developmental disorders
Ethyl alcohol	Male infertility, developmental disorders, birth defects, low birth weight, or premature births
Lead	Infertility, miscarriage, growth retardation, functional deficit, and breast milk contamination
Mercury	Male infertility, birth defects, growth retardation, and breast milk contamination
Phenol	Altered sex ratio, spontaneous abortions, and impotence
Polychlorinated biphenyls (PCBs)	Infertility, spontaneous abortion, growth retardation, and breast milk contamination.
Warfarin	Birth defects, developmental disorders, and spontaneous abortions
Toluene (methyl benzene)	Low birth weight, developmental disorders, birth defects, menstrual disorders, and male and female infertility
Bio-hazardous material	
Cytomegalovirus	Spontaneous abortion, birth defects, growth retardation, and developmental disorders
Hepatitis B virus	Growth retardation, liver disease in infected offspring, and breast milk contamination
HIV	Functional deficit and childhood cancer
Rubella virus (German measles)	Birth defects, growth retardation, and developmental disorders
Varicella-zoster virus (chicken pox and shingles)	Birth defects and growth retardation
Physical hazard	
Excessive heat	Male infertility
Heavy physical exertion	Spontaneous abortion and growth retardation
Ionizing radiation	Male and female infertility, spontaneous abortion, birth defects, growth retardation, developmental disorders, and childhood cancer

Table 1.
Category I: Satisfactory reports with human evidence [15].

2.2.2 Category II: satisfactory reports with animal evidence/limited human evidence

The toxicants which are likely found to be or possible human reproductive hazards are considered as category-II agents. In order to support this category, studies which include experimental animals and/or limited human trails can be considered.

Chemical	Reported adverse effects
Acetaldehyde	Growth retardation and developmental disorders
Acetone	Female infertility, birth defects, and menstrual disorders
Aluminum	Birth defects
Ammonia	Premature birth
Anesthetic agents	Male infertility, spontaneous abortion, birth defects, growth retardation, and breast milk contamination
Antimony	Spontaneous abortion, and breast milk contamination
Arsenic	Birth defects and spontaneous abortion
Benzene	Female infertility, spontaneous abortion, birth defects, growth retardation, and menstrual disorders
Boric acid, borates	Reduced male sex drive, male infertility, and female infertility
Bromine	Male infertility, decreased libido, impotence, and breast milk contamination
Cadmium	Infertility, birth defects, growth retardation, developmental disorders, and breast milk contamination
Carbamide (urea)	Spontaneous abortion
Carbaryl	Male and female infertility and genetic defects
Carbon tetrachloride	Male and female infertility
Chloroform	Spontaneous abortion and birth defects
Copper	Spontaneous abortion and birth defects
Dimethoate	Birth defects, spontaneous abortion, and male infertility
Dimethylformamide, N, N (DMF)	Spontaneous abortion, stillbirths, birth defects, and female infertility
Ethylene glycol monomethyl ether (EGME)	Male infertility, birth defects, and developmental disorders
Ethylene oxide	Male and female infertility, spontaneous abortion, birth defects, and growth retardation
Formaldehyde	Female infertility and spontaneous abortion
Gasoline	Female infertility, birth defects, and menstrual disorders
Lithium	Birth defects and male infertility among patients taking lithium
Manganese	Reduced male sex drive, male infertility, and breast milk contamination
Nitrous oxide	Male and female infertility, spontaneous abortion, and developmental defects
Oral contraceptives	Reduced male sex drive, female infertility, and birth defects
Paints	Spontaneous abortion and developmental disorders
Polyvinyl chloride (PVC resin)	Female infertility, spontaneous abortion, and stillbirths
Solvents	Birth defects, developmental disorders, spontaneous abortion, impotence, female infertility, menstrual disorders, and breast milk contamination
Sulfur dioxide	Spontaneous abortions, female infertility, low fetal weights, and birth defects
Styrene (vinyl benzene)	Male and female infertility, spontaneous abortion, and breast milk contamination

Chemical	Reported adverse effects
Tetrachloroethylene (perchloroethylene)	Female infertility, spontaneous abortion, developmental disorders, birth defects, menstrual disorders, and breast milk contamination
Trichloroethylene	Male and female infertility, spontaneous abortion, and birth defects
Trinitrotoluene	Male infertility
Vinyl chloride monomer	Reduced male sex drive, spontaneous abortion, birth defects, and childhood cancer
Xylene	Female infertility, birth defects, menstrual disorders, and breast milk contamination
Physical hazard	
Low atmospheric pressure (hypobaric)	Male infertility and growth retardation
High atmospheric pressure (hyperbaric)	Male infertility and birth defects

Table 2.
Category II: Satisfactory reports with animal evidence/limited human evidence [15].

Chemical	Reported adverse effects
Acrylamide	Male and female infertility, birth defects, and developmental disorders
Carbon dioxide	Birth defects and male infertility
Carbon tetrachloride	Male and female infertility, developmental disorders, and birth defects
Chromium	Birth defects and infertility
Dimethyl phthalate	Birth defects and developmental disorders
Dimethyl sulfoxide (DMSO)	Developmental disorders
Epichlorohydrin	Male infertility
Ethylene thiourea	Birth defects
Halothane	Developmental disorders and birth defects
Methyl alcohol	Developmental disorders
Methyl ethyl ketone (MEK)	Developmental disorders
Methylformamide, N	Birth defects
Methylpyrrolidone	Birth defects
Nickel	Birth defects
Polybrominated biphenyls (PBBs)	Birth defects and developmental disorders
Ribavirin (virazole)	Birth defects and spontaneous abortion
Toxaphene (camphechlor)	Developmental disorders, infertility, and breast milk contamination
1,1,1-Trichloroethane	Low fetal weight, birth defects, and developmental disorders

Table 3.
Category III: Suspect/insufficient reports with animal evidence but not humans [15].

To support this class, minimum criteria is a single, systematic experiment on one animal species for one adverse reproductive effect. Below **Table 2**, enlisted the toxic effects of potential reproductive toxicants based on the observation of studies comprised animals and humans.

2.2.3 *Category III: suspect/insufficient reports with animal evidence but not humans*

This category consists of agents with probable or indeterminate reproductive hazards. Though they possess adverse effect on reproductive health but data are inadequate. Present details in **Table 3**, is only of studies with animal experiments with no human trials.

3. Chemical factors

Chemicals are omnipresent elements with both positive and negative effects found in workplaces across the globe. Several environmental chemicals together with other agents (e.g., radiation and bacteria), chemicals may also destructively affect the reproductive systems of male and female workers (**Table 4**). Exposure to toxicants before and after conception can affect parents, fetuses, and newborns. In most of the working environments, huge numbers of workers are exposed to the substances which are potentially toxic to reproductive health even after knowing. Exposure to industrial chemicals can alter reproductive functions in females. The ovary of a female is vulnerable in most of the cases therefore, have a significant effect on fertility, menstrual (estrous) cyclicity, and the timing of puberty and menopause [26]. Many toxic chemical agents, active metabolites from mother may reach the womb by different routes of mechanisms causing unfavorable environment to its development. This toxicity could reach to the deepest point, where it may not only obstructs the transport of male and female gametes to the site of fertilization but also stops fertilized egg moving to the site of implantation and development in the uterus. It is also found that, abnormal hormonal control during pregnancy is influenced by toxicants resulting in potential adverse effects on the fetus. The primary manifestations of developmental toxicity are embryo/fetal death, malformations (birth defects), growth retardation, and developmental delay. Adverse fetal outcomes may also include preterm delivery, altered sex ratio, and childhood cancer. The human tests is the house of high rates of proliferation, differentiation, as well as a metabolic activity associated with the production of large quantities of mature sperm which makes it more vulnerable to chemicals. The toxicants will target the Leydig cells (LC), sertoli, and germ cells of a testis which are the site of spermatogenesis, leading to germ-cell apoptosis and spermatogenic failure. Examples of chemicals toxic to the male reproductive system are presented in **Table 5**.

3.1 Heavy metals

Metals exert an extensive diversity of hazardous effects on reproduction and development including influence on fertility, intrauterine growth retardation, abortions, malformations, birth defects, and developmental effects, mainly those on the nervous system [21]. More recent, important mechanisms of action are those related to endocrine disruption and oxidative stress. Endocrine disruptors (EDs) have been defined as "exogenous chemical substances or mixtures able to alter the structure or function of the endocrine system and to cause adverse effects on organisms or their progeny" [28].

It is believed that, partial exposure to certain chemicals will decreases the puberty, causes abnormal semen quality and quantity, impairment of sex ratio, occurrence of hypospadias, testicular cancer, infertility, miscarriages, and genetic defects. In an *in vitro* study, concentration over 1 mmol of copper significantly

Agent	Industry or occupational group	Reported effects of female exposure	Reported effects of male exposure
Organic solvents in general	Painting, degreasing, shoemaking, printing, dry cleaning, metal industry, and several other fields of industry	Reduced fertility, menstrual disorders, fetal loss, birth defects, preterm birth, neurobehavioral effects, and childhood leukemia	Delayed conception, reduced semen quality, fetal loss, and birth defects
Benzene	Petrochemical industry and laboratory personnel	Fetal loss, reduced fertility, and low birth weight	
Carbon disulfide	Viscose rayon industry	Menstrual disorders	Decreased libido and potency
Some ethylene glycol ethers and their acetates	Electronics industry, silk screen printing, photography and dyeing, shipyard painting, metal casting, chemical industry, and other industries	Reduced fertility, fetal loss, birth defects, and menstrual disorders	Reduced semen quality
Tetrachloroethylene	Dry cleaning and degreasing	Reduced fertility and fetal loss	
Toluene	Shoe industry, painting, and laboratory work	Reduced fertility and fetal loss	
Metals			
Lead	Battery industry, lead smelting, foundries, pottery industry, ammunition industry, and some other metal industries	Reduced fertility, fetal loss, preterm birth, low birth weight, birth defects, and impaired cognitive development	Reduced semen quality, reduced fertility, fetal loss, and birth defects
Inorganic mercury	Lamp industry, chloralkali industry, and dental personnel	Reduced fertility, menstrual disorders, and fetal loss	Fetal loss
Pesticides[a]	Agriculture, gardening, and greenhouse work	Reduced fertility, fetal loss, birth defects, preterm birth, reduced fetal growth, neurodevelopmental effects, and childhood leukemia	Reduced sperm quality, reduced fertility, fetal loss, birth defects, and childhood cancer
Pharmaceuticals			
Anesthetic gases	Operating rooms, delivery wards, and dental offices	Fetal loss, reduced birth weight, preterm birth, birth defects, and reduced fertility	
Nitrous oxide	Operating rooms, delivery wards, and dental offices	Fetal loss, reduced birth weight, and reduced fertility	

Agent	Industry or occupational group	Reported effects of female exposure	Reported effects of male exposure
Antineoplastic agents	Hospital workers, pharmaceutical industry	Menstrual dysfunction, reduced fertility, fetal loss, premature birth, low birth weight, and birth defects	
Carbon monoxide	Iron and steel foundries, welding, food industry, car repair, and service stations	Preterm birth and intrauterine death	

[a]Examples of pesticides with adverse effects in men include dibromochloropropane (DBCP), 2,4-dichlorophenoxyacetic acid (2,4-D), ethylene dibromide, chlordecone, carbaryl, alachlor, atrazine, and diazinon.

Table 4.
Adverse reproductive effects of some chemical agents that have been reported in human studies (Source: [25]).

reduced the human sperm motility, where sperm motility is considered as one of the prime attribute of a male gamete to reach oocyte [29]. Cadmium, for example, may affect steroidogenesis by mimicking or inhibiting the actions of endogenous estrogens [30]. Several metals such as, iron, copper, cobalt, and lead will lead to oxidative stress by increasing the production of reactive oxygen species, decrease the levels of glutathione and other antioxidants. Lead interrupts the hypothalamic-pituitary axis and has been reported to alter hormone levels [31, 32], alter the onset of puberty, and decrease overall fertility [31]. The industrial discharges and emissions, batteries and most of the thermometers are the primary sources of mercury. Currently, mercury concentrations can be found in food chain especially in tainted seafood, leading to bioaccumulation amongst humans who are consumers of such foods which leads to reproductive toxicity [31] by disrupting normal spermatogenesis and fetal development [32]. Amongst heavy metals, boron is widely employed in the production of soap and cement including in leather industries, which is found to disrupt the HPG axis like lead [32]. Cadmium is another metal which is reported to cause testicular necrosis in mice and notable libido activity and infertility (**Table 5**) [33].

3.2 Insecticides

Insecticides are described as "chemicals used to control insects by killing them or preventing them from engaging in undesirable or destructive behaviors" by United States Environmental Protection Agency [34]. Considering the chemical structure, insecticides could be divided into five groups: (i) organochlorines, (ii) organophosphates, (iii) carbamates, (iv) pyrethrins/pyrethroids, and (v) nicotine/neonicotinoids. Insecticides could be characterized as "endocrine disrupters" due to their adverse effects on reproductive hormone pathway [35]. Exposure to permethrin, fenvalerate, and cypermethrin showed drastic drop in serum testosterone levels and elevated follicle stimulating hormone (FSH) and luteinizing hormone (LH) levels. It is understandable that, decrease in testosterone levels provides a negative feedback to HPG axis, helping FS and LH to rise. Impairment of spermatogenesis, anti-androgenic effects, and alterations in reproductive enzyme pathways, decreased sperm quality and motility are key elements in insecticide-induced male infertility [36]. Organophosphates could alter the spermatozoon chromatin structure, DNA, acrosome, motility and, have toxic effects on HPG axis

Chemical hazards	Species effect observed (h = humans, a = animals)	Examples of occupations where hazards may occur
Alcohol	h	Social threat
Alkylating agents	h, a	Chemical and drug industrial
Anesthetic gases nitrous oxide	a, h	Medical, dental, and veterinary employees
Cadmium	h, a	Storage batteries and smelter workers
Carbon disulfide	h, a	Viscose rayon manufacture and soil treaters
Carbon tetrachloride	a	Chemical laboratories and dry cleaners
Diethylstilbestrol (DES)	a, h	DES producers
Chloroprene	h, a	Rubber labors
Ethylene oxide	a, h	Health-care workers (disinfectants) and users of epoxy resins
Hair dyes	a	Cosmetic manufacturers, hairdressers, and barbers
Lead	h, a	Storage batteries, policeman, and smelter workers
Manganese	h	Welders, ore smelters, and roasters
Nickel	a	Smelters and welders
Organic mercury compounds	a	Pesticide workers
Tris (flame retardants)	a, h	Clothing and textile work
Pesticides	a, h	Farmworkers and pesticide manufacturers
Dibromochloropropane	—	Exterminators
Vinyl chloride	h	Polyvinylchloride manufacture and processing
Elevated carbon dioxide	a	Brewery workers and chemical manufacture
Elevated temperature	h, a	Bakers, glassblowers, foundry, and oven workers
Microwaves	h, a	Radar operators, flight crew or pilots, and transmitter operators
X-irradiation	h, a	Health workers and radiation workers

Table 5.
Examples of agents toxic to the male reproductive system [27].

including reduced levels of testosterone [37]. Organochlorines such as dichloro-diphenyltrichloroethane (DDT), methoxychlor, chlordane, heptachlor, aldrin, endrin, and lindane are reported to cause oxidative stress within epididymis by decreasing antioxidant defense [38]. Endosulfan was found to cause irregular sperm maturation amongst farmers who employed this chemical into their fields. These insecticides not only hazard to male but also to female reproductive system (**Table 6**). They disrupt endocrine system and ovarian physiology in females via HPG axis leading to follicular maturation anomaly, disordered ovarian cycle, prolonged pregnancy, stillbirth, and subfertility also DNA damage and apoptosis amongst cells [39]. Endosulfan, an organochlorine, triggered apoptosis via

Hazard	Outcome
Anesthetic gases	Miscarriage and neonatal deaths
Hepatitis B	Newborn hepatitis and liver cancer
Organic mercury	Cerebral palsy and brain malformation
Lead	Abortions and premature birth (polychlorinated biphenyls)
Radiation	Miscarriage, brain defects, and skeletal defects [suspected reproductive hazards (based on human studies)]
Carbon monoxide	Slowed growth
Cytotoxic drugs	Abortions
Ethylene oxide	Abortions
Hexachlorophene	Birth defects
Organic solvents	Cleft palate, miscarriage, newborn infection, childhood cancer, physical stress (including heat), and prematurity
Vinyl chloride	Brain defects

Table 6.
Examples of reproductive hazards to humans.

oxidative stress induction in the follicle cells. Moreover, it induced the expressions of steroidogenic acute regulatory protein (StAR), CYP19A1a and aromatase, causing improper ovarian maturation organochlorines (**Table** 7) [38].

3.3 Genital tract infection

The male accessory gland infection and genital tract infections, by numerous bacteria, viruses, and fungi have an adverse effect on male fertility aptitude by infecting semen, inducing oxidative stress, which damages testicles, colonizing genital tract leading to obstruction, and directly disturbing sperm function, morphology [40]. The sexually transmitted pathogens and uro-pathogens such as, *Chlamydia trachomatis*, *Escherichia coli*, *Staphylococcus epidermidis*, *Klebsiella*

Insecticides	Effects on endocrine system
Aldicarb	17 beta-estradiol and progesterone inhibition
Aldrin	Androgen receptor binding
Carbofuran	Estradiol and progesterone increase; testosterone decrease (chlordane)
Deltamethrin	Estrogenic activity
Dieldrin	Androgen receptor binding, inducing estrogen receptor production in the cell (endosulfan)
Lindane	Luteal progesterone decrease, androgen, estrogen, and progesterone receptor binding
Methoxychlor	Estrogenic effect, pregnane X cellular receptor binding (parathion)
Parathion	Gonadotrophic hormone synthesis inhibition
Fenoxycarb	Testosterone metabolism disruption
Endosulfan	Androgen receptor binding, inducing estrogen receptor production in the cell
Chlordane	Androgen receptor binding, estrogenic pathway inhibition

Table 7.
Selected insecticides and their effects on endocrine system.

spp., *Proteus* spp., *Ureaplasma urealyticum*, *mycoplasms*, *Trichomonas vaginalis*, *Staphylococcus saprophyticus*, *Neisseria gonorrhoeae*, and chronic viral sperm contagions *viz. human immunodeficiency virus*, *hepatitis B*, and *hepatitis C viruses* [41] are some of the prominent pathogens which infect the male accessory glands and also the genital tract leading to male infertility which is accounted for 15–20% of total infertility [42]. The testicular tenderness, urethral expulsion, epididymal inflammation, and throbbing ejaculation are some of the notable symptoms of genital infection [43]. Microbial infection triggers the immune inflammatory mechanism, by which white blood cells and pro-inflammatory cytokines such as IL-1, IL-6, IL-8, macrophages, and polymorphonuclear neutrophils will be released into the infection site, which have been found to show negative effect on sperm functionality and fertilizing ability [44]. In females, genital tract infection can lead to adverse health outcomes such as infertility, ectopic pregnancy and increased vulnerability to transmission of the human immunodeficiency virus. It is also associated with adverse pregnancy outcomes. Vaginitis and cervicitis are common lower genital tract infections usually found in females which enable uncharacteristic vaginal discharge, genital discomfort, itching, and burning sensation while urination. Generally infection occurs at soft tissue and perineal of female genital tract. The infections which are common *viz.*, bacterial vaginosis, Bartholin gland abscess, endometritis, *pyometra*, *salpingitis*, pelvic inflammatory disease, intrauterine contraceptive device-associated infection, postsurgical obstetric, and gynecologic infections.

3.4 Obesity

A medical ailment linked with excessive accumulation of white adipose tissue within the body, distressing normal health and a person with BMI 25–30 kg/m^2 can be overweight, whereas BMI ≥30 kg/m^2 is said to be obese [45]. Current evidences have shown the destructive impact of obesity on the reproductive aptitude of men by subduing spermatogenesis, causing abnormal sperm morphology, sperm DNA fragmentation, erectile dysfunction, and reduced libido [46]. Increased deposition of fat in the upper thighs, scrotum, and suprapubic area causes rise in scrotal temperature and oxidative stress, which ruins normal spermatogenesis, sperm motility, and also interferes with sperm-oocyte interaction [47]. Prevalence of menstrual dysfunction, subfertility, abortion rates, pregnancy hitches, and anovulation are commonly seen in overweighed women and they are at high risk for reproductive health. In obese women, gonadotropin secretion is affected because of the increased peripheral aromatization of androgens to estrogens. When neuro-regulation of HPG axis declines abnormal ovulatory activities can be seen [48], which is generally because of decreased sex hormone-binding globulin (SHBG), growth hormone (GH), and insulin-like growth factor binding proteins (IGFBP) leading to elevated leptin levels [49].

3.5 Tobacco consumption and smoking

Chewing tobacco and smoking are the injurious addictions [50], which contains >30 mutagenic substances, numerous toxic chemicals along with nicotine and familiar carcinogens [51], have been reported for adversely affecting semen quality and eventually male infertility [52]. The cytotoxic effect of tobacco chewing and/or smoking decreases sperm count, motility, viability and morphological mutations along with damaging testes, accessory glands/ducts leading to low semen volume, seminal leukocytes, abnormal hormonal levels, impaired spermatogenesis, sperm DNA damage, oxidative stress, cytogenetic abnormalities, spontaneous abortions, and congenital anomalies [53]. For women, smoking cigarettes can lead to reproductive damage, reduced fertility, and difficulty conceiving. Research shows smoking may affect hormone production, making it difficult to become pregnant [54]. Several studies have indicated the adverse effects of maternal smoking during pregnancy on abnormal fetus development, newborn deaths, and problems associated with pregnancy resulting in premature conception.

3.6 Alcohol ingestion

Chronic and excessive alcohol consumption hamper the normal functioning of the HPG axis, resulting abnormal secretion of gonadotropin-releasing hormone (GnRH), FSH, LH, and testosterone that alters LCs and sertoli cell functions, and impairs spermatogenesis [55]. Furthermore, prolonged alcohol addiction causes testicular damage and shrinkage [56], abridged semen quality [57], lower semen volume [58], partial or complete spermatogenic seizure [59], and delayed seminal fluid liquefaction [60]. Eventually, decline in sex hormone levels causes loss of secondary sexual distinctiveness, Sertoli cell-only syndrome, impotence, diminished libido [61], erectile dysfunction, and ejaculation problems [62]. Heavy alcohol use may diminish ovarian reserve and fecundability in women. Detrimental effects of mild to moderate alcohol consumption may interfere with normal menstrual cycle, disturb puberty, damage reproductive capacity, and cause hormonal abnormality amongst women [63]. As alcohol easily pass through placenta, accumulates in amniotic fluid leading to decreased fetal metabolic enzyme activity [64].

3.7 Drugs

Drugs of abuse and chronic medication may have adverse effect on the fertility potential of men by disturbing HPG axis, gonadotoxic activity, or by upsetting sexual performances (ejaculation, erection, and libido) [65]. Prolonged treatment with immunosuppressive drugs (sirolimus and ciclosporine), corticosteroids, immunomodulators (mAbs and TNFα inhibitors), thyrosine kinases inhibitors, opiates (morphine and cocaine), hormonal agents (anabolic steroids and testosterone), antiandrogenic drugs (cyproterone acetate and flutamide), antibiotics (erythromycine and tetracyclines), antimicrobial drugs (metronidazole and chloroquine), antidepressant (imipramine and buspirone), antipsychotic (phenothiazines and butyrophenones), etc., will present a drastic drop in sperm count, motility and morphology [66], inhibition or low level of testosterone [67], hindering acrosomal reaction and shrinking fertility potential of spermatozoa [68], toxic effect on gonads [69], drop in testicular size, weight and volume, inhibiting dopamine synthesis [70] thereby causing erectile dysfunction [71], decreased libido [72], sedation [73] and delayed ejaculation [74], anejaculation/retrograde ejaculation [75] which will result in impotency or male infertility (**Tables 8** and **9**) [65].

3.8 Testicular hyperthermia

The normal spermatogenesis in humans and most mammals require testicular temperature 2–4°C below body temperature [77]. The rise in the scrotal temperature and its duration upsets semen quality resulting spermatogenesis seizure [78], producing more morphologically abnormal sperm, reduced sperm movement [79], destruction of mitochondria and DNA [80], declined sperm concentration [81], and death of germ cells [82], sooner or later into male infertility. The relentless exposure to several external factors such as stance/posture, outfit/clothing, lifestyle, and seasons may negotiate the ability of the scrotum to thermo-regulate leading to adverse effects on male fertility [83]. Apart from these factors, occupational exposure to

Medication	Effect on reproductive function
Anabolic steroids	Impairment of spermatogenesis (up to 1 year recovery); may cause hypogonadism through pituitary-gonadal axis
	Reversible
Antiandrogens	Impairment of spermatogenesis; erectile dysfunction
Cyproterone acetate, danazol, finasteride, ketoconazole, and spironolactone	Reversible
Antibiotics	Impairment of spermatogenesis
Ampicillin, cephalotin, cotrimoxazole, gentamycin, neomycin, nitrofurantoin, Penicillin G, and spiramycin	Reversible
Antibiotics	Impairment of sperm motility
Cotrimoxazole, dicloxacillin, erythromycin, lincomycin, neomycin, nitrofurantoin, quinolones, tetracycline, and tylosin	Reversible
Antiepiletics	Impairment of sperm motility
Phenytoin	Reversible
Antihypertensives	Fertilization failure
Antipsychotics	Impairment of spermatogenesis and sperm motility
Butyrophenones	Reversible
H2 blockers: cimetidine, ranitidine	Increase prolactin concentrations that can lead to impairment of luteal function, loss of libido, and erectile dysfunction

Table 8.
Medications and their respective effects on both male and female reproductive function [76].

Chemical	Possible reproductive effects
BPA	Inhibits binding to androgen receptor, decreased semen quality, erectile dysfunction, chromosomal abnormalities in oocyte, and recurrent miscarriage
Disinfection by-products	
Organochemicals and pesticides e.g., DDT, DDE, methoxyclor	Change in hormone levels, irregular menstruation, decreased fertility, decreased semen quality, chromosomal abnormalities in sperm, altered histology of testes, decreased libido, fetal loss, and miscarriage
Dioxins	Changes in hormone levels, altered puberty, altered start of menarche, endometriosis, decreased fertility, and fetal loss
Phthalates	Decreased semen quality, oligozoospermia, earlier menarche, altered menstrual cycle, and infertility
Solvents	Change in hormone levels, decreased semen quality, irregular menstruation, decreased fertility, miscarriage, and fetal loss

Table 9.
Chemicals and their respective effects on both male and female reproductive function [76].

extreme temperature, for example, workers at welding factories, ceramics companies, furnace workers, mechanics, and drivers are the chief victims of this risk factor facing fertility problems [47].

4. Conclusion

The male or female reproductive structure or function disturbed by any factors leading to the delivery of abnormal offspring, which has interfered with the continuation of generation is basically a reproductive or developmental toxicity. Presently, several reproductive or developmental toxicants are under routine by the people without their awareness, which obviously have negative impact on their health. In most of the working environments, due to the lack of knowledge and information many workers are occupationally exposed to such hazards and they are at the edge of reproductive toxicity. To understand the pathway of this toxicity needs a deeper research but due complexity of the mammalian reproductive cycle *in vitro* studies are quite lagging but one can slice this series of cycle and work on it independently. Currently, advancement in the field of reproductive toxicity testing has come-up with useful and promising *in vitro* models but their potential and accuracy are yet to be finalize. Though, individual tests are helping to identify certain aspects of toxicity but study can be only completed with combination of detailed toxicology reports.

Author details

Makhadumsab Toragall[1], Shridhar C. Ghagane[2]*, Rajendra B. Nerli[3] and Murigendra B. Hiremath[4]

1 Department of Biotechnology and Microbiology, Karnatak University, Dharwad, Karnataka, India

2 KAHER's Dr. Prabhakar Kore Basic and Applied Science Research Center (BSARC), V. K. Institute of Dental Sciences, Belagavi, Karnataka, India

3 Division of Urologic-Oncology, Urinary Biomarkers Research Centre, KLES Dr. Prabhakar Kore Hospital and Medical Research Centre, Belagavi, Karnataka, India

4 Department of Urology, JN Medical College, KLE Academy of Higher Education and Research (Deemed-to-be-University), Belagavi, Karnataka, India

*Address all correspondence to: shridhar.kleskf@gmail.com

IntechOpen

© 2021 The Author(s). Licensee IntechOpen. This chapter is distributed under the terms of the Creative Commons Attribution License (http://creativecommons.org/licenses/by/3.0), which permits unrestricted use, distribution, and reproduction in any medium, provided the original work is properly cited.

References

[1] Balme DM, editor. Aristotle History of Animals: Books VII-X. Cambridge, Massachusetts, United States: Harvard University Press; 1991

[2] Johnston DR. The history of human infertility. Fertility and Sterility. 1963; **14**(3):261-272

[3] Behjati-Ardakani Z, Akhondi MM, Mahmoodzadeh H, Hosseini SH. An evaluation of the historical importance of fertility and its reflection in ancient mythology. Journal of Reproduction and Infertility. 2016;**17**(1):2

[4] Sattari J. Myth in Today's World. Tehran: Markaz Publication; 2004

[5] Hofmann T. Reproductive toxicology. In: Vogel HG, Hock FJ, Maas J, Mayer D, editors. Drug Discovery and Evaluation: Safety and Pharmacokinetic Assays; with 125 Tables. Berlin, Germany: Springer Science & Business Media; 2006. pp. 1317-1327

[6] Kumar N, Singh AK. Trends of male factor infertility, an important cause of infertility: A review of literature. Journal of Human Reproductive Sciences. 2015;**8**(4):191-196

[7] Masoumi SZ, Parsa P, Darvish N, Mokhtari S, Yavangi M, Roshanaei G. An epidemiologic survey on the causes of infertility in patients referred to infertility center in Fatemieh Hospital in Hamadan. Iranian Journal of Reproductive Medicine. 2015;**13**(8): 513-516

[8] Bremer S, Cortvrindt R, Daston G, Eletti B, Mantovani A, Maranghi F, et al. Reproductive and developmental toxicity. Alternatives to Laboratory Animals. 2005;**33**(1_suppl):183-209

[9] Wilson JG. Embryological considerations in teratology. In: Wilson JG, Warkany J, editors. Teratology, Principles and Techniques. Chicago: The University of Chicago Press; 1965. pp. 251-277

[10] Food and Drug Administration. Guideline for Reproduction Studies for Safety Evaluation of Drugs for Human Use. USA: Food and Drug Administration; 1966

[11] Committee on Safety of Medicines. Guidelines on Reproduction Studies for Guidance of Applicants for Product Licences and Clinical Trial Certificates. UK: Committee on Safety of Medicines; 1974

[12] Ministry of Health and Welfare. Guidelines for Reproduction Studies for Safety Evaluation of Drugs. Japan: Ministry of Health and Welfare; 1975

[13] Bass R, Ulbrich B, Neubert D, Omori Y, Palmer A. Draft guideline on detection of toxicity to reproduction for medicinal products. Adverse Drug Reactions and Acute Poisoning Reviews. 1991;**10**(3):127-141

[14] ICH Harmonized Tripartite Guideline. Detection of toxicity to reproduction for medicinal products. In: Endorsed by the ICH Steering Committee at Step 4 of the ICH Process. 1993

[15] UNECE. Globally harmonized system of classification and labelling of chemicals (GHS). United Nations Economic Commission for Europe; 2003 [Accessed: 23 May 2021]

[16] Food and Drug Administration. Guidance for "Industry Reproductive and Developmental Toxicities—Integrating Study Results to Assess Concerns". Silver Spring, MD: Food and Drug Administration; 2011

[17] Osha U. Globally Harmonized System of Classification and Labelling

of Chemicals (GHS). United states: United Nations Economic Commission for Europe; 2013. p. 224

[18] USEPA. Guidelines for developmental toxicity risk assessment. Federal Register. 1991;**56**(234): 63798-63826

[19] Mattison DR, Thomford PJ. The mechanisms of action of reproductive toxicants. Toxicologic Pathology. 1989;**17**(2):364-376

[20] Levine H, Jorgensen N, Martino-Andrade A, Mendiola J, Weksler-Derri D, Mindlis I, et al. Temporal trends in sperm count: A systematic review and meta-regression analysis. Human Reproduction Update. 2017;**23**(6):646-659

[21] Apostoli P, Telišman S, Sager PR. Reproductive and developmental toxicity of metals. Handbook on the Toxicology of Metals. 2007;**1**:213-249

[22] Crain DA, Janssen SJ, Edwards TM, Heindel J, Ho SM, Hunt P, et al. Female reproductive disorders: The roles of endocrine-disrupting compounds and developmental timing. Fertility and Sterility. 2008;**90**(4):911-940

[23] Jensen TK, Sobotka T, Hansen MA, Pedersen AT, Lutz W, Skakkebæk NE. Declining trends in conception rates in recent birth cohorts of native Danish women: A possible role of deteriorating male reproductive health. Obstetrical and Gynecological Survey. 2009; **64**(4):246-247

[24] Hamilton BE, Ventura SJ. Fertility and abortion rates in the United States, 1960-2002. International Journal of Andrology. 2006;**29**:34-45

[25] Lindbohm ML, Sallmen M. Reproductive Effects Caused by Chemical and Biological Agents. Helsinki, Finland: Finnish Institute of Occupational Health; 2017

[26] Rim KT. Reproductive toxic chemicals at work and efforts to protect workers' health: A literature review. Safety and Health at Work. 2017;**8**(2): 143-150

[27] Rom WN, Markowitz SB, editors. Environmental and Occupational Medicine. Philadelphia: Lippincott Williams & Wilkins; 2007

[28] Medical Research Council. IEH Assessment on Environmental Oestrogens: Consequences to Human Health and Wildlife. Leicester, England: University of Leicester; 1995. p. 105

[29] Leroy-Martin B, Saint-Pol P, Hermand E. Copper—A major contraceptive agent? Contraception, Fertilité, Sexualité (1992). 1987; **15**:599-602

[30] Henson MC, Chedrese PJ. Endocrine disruption by cadmium, a common environmental toxicant with paradoxical effects on reproduction. Experimental Biology and Medicine. 2004;**229**(5):383-392

[31] Chalupka S, Chalupka AN. The impact of environmental and occupational exposures on reproductive health. Journal of Obstetric, Gynecologic, and Neonatal Nursing. 2010;**39**:84-102

[32] Sikka SC, Wang R. Endocrine disruptors and estrogenic effects on male reproductive axis. Asian Journal of Andrology. 2008;**10**:134-145

[33] King LM, Anderson MB, Sikka SC, George WJ. Murine strain differences and the effects of zinc on cadmium concentrations in tissues after acute cadmium exposure. Archives of Toxicology. 1998;**72**:650-655

[34] EPA. Insecticides, CADDIS Volume 2 by United States Environmental Protection Agency [Internet]. 2020.

Available from: https://www.epa.gov/caddis-vol2/insecticides#main-content [Accessed: 8 April 2020]

[35] Sweeney T, Nicol L, Roche JF, Brooks AN. Maternal exposure to octylphenol suppresses ovine fetal follicle-stimulating hormone secretion, testis size, and sertoli cell number. Endocrinology. 2000;**141**(7):2667-2673

[36] Ghuman SPS, Ratnakaran U, Bedi JS, Gill JPS. Impact of pesticide residues on fertility of dairy animals: A review. The Indian Journal of Animal Sciences. 2013;**83**:1243-1255

[37] Mitra A, Maitra SK. Reproductive toxicity of organophosphate pesticides. Annals of Clinical Toxicology. 2018; **1**:1004-1012

[38] Kara M, Oztaş E. Reproductive toxicity of insecticides. In: Animal Reproduction in Veterinary Medicine. Istanbul, Turkey: IntechOpen; 2020

[39] Sharma RK, Singh P, Setia A, Sharma AK. Insecticides and ovarian functions. Environmental and Molecular Mutagenesis. 2020;**61**:369-392

[40] Ma W, Li S, Ma S, Jia L, Zhang F, Zhang Y, et al. Zika virus causes testis damage and leads to male infertility in mice. Cell. 2016;**167**(6):1511-1524

[41] Lorusso F, Palmisano M, Chironna M, Vacca M, Masciandaro P, Bassi E, et al. Impact of chronic viral diseases on semen parameters. Andrologia. 2010;**42**(2):121-126

[42] Weidner W, Pilatz A, Diemer T, Schuppe HC, Rusz A, Wagenlehner F. Male urogenital infections: Impact of infection and inflammation on ejaculate parameters. World Journal of Urology. 2013;**31**(4):717-723

[43] Dieterle S. Urogenital infections in reproductive medicine. Andrologia. 2008;**40**(2):117-119

[44] Sanocka D, Kurpisz M. Reactive oxygen species and sperm cells. Reproductive Biology and Endocrinology. 2004;**2**(1):1-7

[45] WHO. Obesity: Preventing and managing the global epidemic. World Health Organization Technical Report Series. 2000;**894**:1-253

[46] Campbell JM, Lane M, Owens JA, Bakos HW. Paternal obesity negatively affects male fertility and assisted reproduction outcomes: A systematic review and meta-analysis. Reproductive Biomedicine Online. 2015;**31**(5): 593-604

[47] Durairajanayagam D, Agarwal A, Ong C. Causes, effects and molecular mechanisms of testicular heat stress. Reproductive Biomedicine Online. 2015;**30**(1):14-27

[48] Parihar M. Obesity and infertility. Reviews in Gynaecological Practice. 2003;**3**(3):120-126

[49] Dag ZO, Dilbaz B. Impact of obesity on infertility in women. Journal of the Turkish-German Gynecological Association. 2015;**16**(2):111

[50] Koskinen LO, Collin O, Bergh A. Cigarette smoke and hypoxia induce acute changes in testicular and celebral microcirculation. Upsala Journal of Medical Sciences. 2000;**105**(3):215-226

[51] Richthoff J, Elzanaty S, Rylander L, Hagmar L, Giwercman A. Association between tobacco exposure and reproductive parameters in adolescent males. International Journal of Andrology. 2008;**31**(1):31-39

[52] Dupont C, Faure C, Daoud F, Gautier B, Czernichow S, Lévy R. Metabolic syndrome and smoking are independent risk factors of male idiopathic infertility. Basic and Clinical Andrology. 2019;**29**(1):1-7

[53] Harlev A, Agarwal A, Gunes SO, Shetty A, du Plessis SS. Smoking and male infertility: An evidence-based review. The World Journal of Men's Health. 2015;**33**(3):143

[54] CDC. Smoking and Reproduction Fact Sheet. Surgeon General's Report on Smoking and Health 50th Anniversary. 2019. Available from: https://www.cdc.gov/tobacco/data_statistics/sgr/50thanniversary/pdfs/fs_smoking_reproduction_508.pdf [Accessed: 8 January 2019]

[55] La Vignera S, Condorelli RA, Balercia G, Vicari E, Calogero AE. Does alcohol have any effect on male reproductive function? A review of literature. Asian Journal of Andrology. 2013;**15**(2):221

[56] Muthusami KR, Chinnaswamy P. Effect of chronic alcoholism on male fertility hormones and semen quality. Fertility and Sterility. 2005;**84**(4):919-924

[57] Ricci E, Al Beitawi S, Cipriani S, Candiani M, Chiaffarino F, Viganò P, et al. Semen quality and alcohol intake: A systematic review and meta-analysis. Reproductive Biomedicine Online. 2017;**34**(1):38-47

[58] Li Y, Lin H, Li Y, Cao J. Association between socio-psycho-behavioral factors and male semen quality: Systematic review and meta-analyses. Fertility and Sterility. 2011;**95**(1):116-123

[59] Gaur DS, Talekar MS, Pathak VP. Alcohol intake and cigarette smoking: Impact of two major lifestyle factors on male fertility. Indian Journal of Pathology and Microbiology. 2010;**53**(1):35

[60] Molnár J, Papp G. Alkohol als möglicher schleimfördernder Faktor im Samen: Alcohol as a possible stimulant of mucous production in the semen. Andrologia. 1973;**5**(2):105-106

[61] Onyije FM. Drug: A major cause of infertility in male. Asian Journal of Medical and Pharmaceutical Researches. 2012;**2**:30-37

[62] Sansone A, Di Dato C, de Angelis C, Menafra D, Pozza C, Pivonello R, et al. Smoke, alcohol and drug addiction and male fertility. Reproductive Biology and Endocrinology. 2018;**16**(1):1-1

[63] Emanuele MA, Wezeman F, Emanuele NV. Alcohol's effects on female reproductive function. Alcohol Research and Health. 2002;**26**(4):274

[64] Gupta KK, Gupta VK, Shirasaka T. An update on fetal alcohol syndrome—Pathogenesis, risks, and treatment. Alcoholism: Clinical and Experimental Research. 2016;**40**(8):1594-1602

[65] Semet M, Paci M, Saïas-Magnan J, Metzler-Guillemain C, Boissier R, Lejeune H, et al. The impact of drugs on male fertility: A review. Andrology. 2017;**5**(4):640-663

[66] Iyer R, Fetterly G, Lugade A, Thanavala Y. Sorafenib: A clinical and pharmacologic review. Expert Opinion on Pharmacotherapy. 2010;**11**(11):1943-1955

[67] Iwasaki M, Fuse H, Katayama T. Histological and endocrinological investigations of cyclosporine effects on the rat testis. Andrologia. 1995;**27**(3):185-189

[68] Leroy C, Rigot JM, Leroy M, Decanter C, Le Mapihan K, Parent AS, et al. Immunosuppressive drugs and fertility. Orphanet Journal of Rare Diseases. 2015;**10**(1):1-5

[69] Grunewald S, Jank A. New systemic agents in dermatology with respect to fertility, pregnancy, and lactation. JDDG. Journal der Deutschen

Dermatologischen Gesellschaft. 2015;**13**(4):277-290

[70] Cicero TJ, Davis LA, LaRegina MC, Meyer ER, Schlegel MS. Chronic opiate exposure in the male rat adversely affects fertility. Pharmacology Biochemistry and Behavior. 2002; **72**(1-2):157-163

[71] Katz N, Mazer NA. The impact of opioids on the endocrine system. The Clinical Journal of Pain. 2009;**25**(2): 170-175

[72] Whirledge S, Cidlowski JA. Glucocorticoids, stress, and fertility. Minerva Endocrinologica. 2010;**35**(2): 109

[73] Yilmaz BA, Konar VA, Kutlu SE, Sandal SU, Canpolat SI, Gezen MR, et al. Influence of chronic morphine exposure on serum LH, FSH, testosterone levels, and body and testicular weights in the developing male rat. Archives of Andrology. 1999;**43**(3):189-196

[74] Bar-Or D, Salottolo KM, Orlando A, Winkler JV, Tramadol ODT Study Group. A randomized double-blind, placebo-controlled multicenter study to evaluate the efficacy and safety of two doses of the tramadol orally disintegrating tablet for the treatment of premature ejaculation within less than 2 minutes. European Urology. 2012;**61**(4):736-743

[75] Samplaski MK, Lo K, Grober E, Jarvi K. Finasteride use in the male infertility population: Effects on semen and hormonal parameters. Fertility and Sterility. 2013;**100**(3):S71-S72

[76] Sharma R, Biedenharn KR, Fedor JM, Agarwal A. Lifestyle factors and reproductive health: Taking control of your fertility. Reproductive Biology and Endocrinology 2013;11(1):1-5

[77] Mieusset R, Bujan L, Mondinat C, Mansat A, Pontonnier F, Grandjean H. Association of scrotal hyperthermia with impaired spermatogenesis in infertile men. Fertility and Sterility. 1987;**48**(6):1006-1011

[78] Setchell BP. The parkes lecture heat and the testis. Reproduction. 1998; **114**(2):179-194

[79] Naz M, Kamal M. Classification, causes, diagnosis and treatment of male infertility: A review. Oriental Pharmacy and Experimental Medicine. 2017;**17**(2): 89-109

[80] Paul C, Murray AA, Spears N, Saunders PT. A single, mild, transient scrotal heat stress causes DNA damage, subfertility and impairs formation of blastocysts in mice. Reproduction. 2008;**136**(1):73

[81] Jensen TK, Bonde JP, Joffe M. The influence of occupational exposure on male reproductive function. Occupational Medicine. 2006;**56**(8): 544-553

[82] Kanter M, Aktas C, Erboga M. Heat stress decreases testicular germ cell proliferation and increases apoptosis in short term: An immunohistochemical and ultrastructural study. Toxicology and Industrial Health. 2013;**29**(2): 99-113

[83] Sharpe RM. Environmental/ lifestyle effects on spermatogenesis. Philosophical Transactions of the Royal Society, B: Biological Sciences. 2010;**365**(1546):1697-1712

Chapter 6

Testicular Histopathology and Spermatogenesis in Mice with Scrotal Heat Stress

Thuan Dang-Cong and Tung Nguyen-Thanh

Abstract

Chronic heat stress-induced testicular damage and function therefore adversely affect their reproduction. Some research shows that heat stress has a negative effect on histopathological features of testicular tissue structure and spermatogenesis. An animal model was used to evaluate the effect of heat stress on testicular histology changes and spermatogenesis. The mouse model of heat stress was established by submerged in a pre-warmed incubator. The testes' tissue was fixed and stained with hematoxylin–eosin (H&E) for quantitative analysis of histopathological alterations and spermatogenesis according to Johnson scoring system. Mice exposed to heat stress exhibited degenerated and disorganized features of spermatogenic epithelium and reduced spermatogenic cells. Heat stress exposure shows a significantly reduced Johnson score compared to the control condition. The percentage of high Johnsen score points was decreased in heat-stress exposure mice, while the ratio of low Johnsen score points was gradually increased. This chapter describes a mouse model for studying the male reproductive system and applies the Johnsen scores system to assess testicular histopathology in the seminiferous tubule cross-section. Collectively, this chapter indicated a negative impact of heat stress on mouse spermatogenesis as well as the human reproductive system.

Keywords: Chronic heat stress, testicular tissue, Johnson scoring system, spermatogenesis

1. Introduction

High environmental temperatures affect the functionality of many biological systems in the human body, such as the circulatory system, integumentary system, and respiratory system, with consequences on male reproductive activity [1].

The testes' temperatures of the most numerous species of mammal are often lower than body temperature to accommodate normal spermatogenesis [2]. An increase in testicular and epididymal temperatures in men and other mammals leads to reduced sperm output, decreased sperm motility, and increases the percentage of sperm with abnormal morphology [3]. Increases in testicular temperature may be detrimental to spermatogenesis and ultimately cause problems infertility [4, 5]. Elevation of scrotal temperature interrupts spermatogenesis, resulting in reduced the number and motility of spermatozoa, fertilization ability of the surviving sperm, and poor fertilization-embryo [4, 6].

A high-temperature environment, although it may be within the physiological limits, still hurts sperm quality. The optimal testicular temperature for spermatogenesis is from 2 to 4°C lower than the body temperature. Each 1°C increase in testicular temperature leads to a 14% decrease in spermatogenesis [7]. Recently, research by Gong et al. showed that high ambient temperature drastically reduces sperm motility through decreased mitochondrial activity and ATP synthesis [8]. Heat-stressed testicular tissue leads to apoptosis via mitochondrial pathways or DNA damage. Besides, damaged sperm in the vas deferens DNA fracture under the influence of high ambient temperature can lead to male infertility [9]. Therefore, heat-stressed is a high-risk factor affecting the testicular structure and reducing sperm quality.

Scrotal heat stress increases oxidation, resulting in disrupted thermoregulation of the testicle such as reduction of production antioxidant enzymes and production of heat shock proteins, increase in apoptosis and cell death [10]. Heat stress reduces the ability to regulate the temperature of the scrotum, causes oxidative stress, thus causing cellular reactions including mitochondrial dysfunction, increases the reactive oxygen species (ROS), and reduces the production of anti-ROS enzymes [6]. Scrotal heat stress causes a variety of mechanisms that occur in the testes including oxidative stress response, heat shock response, cell cycle checkpoints, DNA repair, apoptosis, and cell death [5, 11]. Scrotal heat stress interrupts spermatogenesis in male mice [5]. Spermatogonial germ cells are also influenced by heat stress, resulting in the elimination and absence of them in the seminiferous tubules, Sertoli and Leydig cells are degenerated [6].

Spermatogenesis is associated with many factors, including environmental temperatures that can cause cell and molecular changes, affecting gene expression that disrupts sperm production, resulting in reduced reproductive health [12]. Germinal cells are variably sensitive to high temperatures [1]. Significant apoptotic loss of germ cells after testicular heat stress may occur either through intrinsic or extrinsic pathways [13]. Cells in the testis of mammals are affected by heat stress principally primary spermatocytes and early spermatids [14]. The testes of the mice in the Heat stress group showed degenerative changes with spermatic arrest in most of the seminiferous ducts [15]. DNA fragmentation in sperm was observed in the testicular tissue exposed to heat stress [16]. The biomechanism of sperm DNA damage involves various types of oxidative reactions, DNA repair errors in the late stages of spermatogenesis, and functional abnormalities that reduce the protective ability of Sertoli cells leading to increased DNA fragmented sperms [17].

Many factors contribute to increasing testicular temperature. These factors can be grouped according to habits, lifestyles [18, 19], occupational factors that must be exposed to high temperatures for long periods [20, 21], and climate change [22]. Depending on the situation, the testicles may experience transient or prolonged heat stress with varying intensity. The effects of high ambient temperature on male fertility tend to accumulate with repeated exposure over a while [7, 18].

Mice are animal models commonly used in biomedical research, because they are easy to handle, cost-effective, and have similarities in thermostats such as a human. Several studies have used experimental mouse models to assess environmental stress on testicular structure and spermatogenesis. Previous research has shown that germ cell death can be induced in response to scrotal temperature exposure in mice [23, 24]. However, the animal model for heat stress study is not yet well established.

The aim of this chapter is to describe a mouse model for investigating the testicular histomorphometric change in response to testicular heat stress and discuss the negative effect of chronic scrotal heat stress on the human male reproductive system.

2. Method to generate a mouse model for testicular heat stress

2.1 Mouse model for testicular heat stress

Swiss mature male mice have 8–10 weeks old (20-23 g) were kept in an animal facility at a controlled temperature (25 ± 1°c) and illuminated (12 h light-12 h dark), and with free access to food and water.

The mouse model was established to study the effect of scrotal heat stress on spermatogenesis and male fertility. The lower body (including the scrotum) of the mouse was soaked in a thermally controlled water bath (heat exposure at 37°C, 40°C, or 43°C) 2 times a day 10 minutes apart, every 10 minutes, lasting for 5 consecutive weeks, 6 days a week. Control mice were treated in the same way, but in water bath maintained at room temperature. After having the bath, mice were dried and examined for any injury or redness to the scrotal skin before being returned to their cages. Mice were kept in different cages and had free access to water and food. All animals are cared for under identical environmental conditions and monitored their general health.

2.2 Tissue processing and hematoxylin–eosin staining

After completing the heat exposure experiment after 5 weeks, all mice was sacrificed under anesthesia. Their testis tissue were harvested for histological morphometric analysis. The testicular specimens were individually immersed into 4% buffered formaldehyde and dehydrated with graded concentrations of ethyl alcohols at room temperature. The testicular specimens were then embedded into paraffin. The paraffin blocks were cut thinly with a thickness of 5 μm and transferred into gelatinized slides. The sections were deparaffinized with xylene and then rehydrated through a descending series of ethanol and water. Slides were stained with hematoxylin and eosin (H&E) and observed under a light microscope for histopathological analysis.

2.3 Johnsen's mean testicular biopsy score count

Testicular histological damage and spermatogenesis were assessed using Johnsen's mean testicular biopsy score under light microscopy [25]. Thirty tubules

Score	Description
10	Complete spermatogenesis with many spermatozoa. Germinal epithelium organized in a regular thickness leaving an open lumen.
9	Many spermatozoa are present but germinal epithelium disorganized with marked sloughing or obliteration of lumen.
8	Only a few spermatozoa are present in the section.
7	No spermatozoa but many spermatids present.
6	No spermatozoa and only a few spermatids are present.
5	No spermatozoa, no spermatids but several or many spermatocytes present.
4	Only few spermatocytes and no spermatids or spermatozoa are present.
3	Spermatogonia are the only germ cells present.
2	No germ cells but Sertoli cells present.
1	No cells in the tubular section

Table 1.
Histological classification of seminiferous tubular cross-sections according to the Johnsen scoring system [25].

for each testis were graded and each tubule was given a score from 1 to 10 based on the presence or absence of germ cell types in the testicular seminiferous tubules such as spermatozoa, spermatids, spermatocyte, spermatogonia, germ cells, and Sertoli cells to evaluate histology. A higher Johnsen's score indicates a better status of spermatogenesis, while a lower score refers to more severe dysfunction. A score of 1 means no epithelial maturation is considered for the tubules with complete inactivity while a score of 10 means full epithelial maturation is considered for the tubules with maximum activity (**Table 1**).

3. The impact of heat stress on spermatogenesis

3.1 An overview of the spermatogenesis

Mammalian spermatogenesis is a complex system within structurally well-designed seminiferous tubules of the testes (**Figure 1a**). Germ cells develop in a spatially organized fashion from the basal membrane to the lumen. During the process of spermatogenesis, primary spermatocytes were differentiated from diploid spermatogonia (Spermatogonial stem cells). Primary spermatocytes undergo to the first meiotic division to generate two secondary spermatocytes. Each secondary spermatocytes go through the second meiosis division to produce haploid daughter spermatids. Spermatid transforms into spermatozoa by the process of metamorphosis (spermiogenesis). Finally, mature sperm were released from the seminiferous epithelium into the tubular lumen. In humans, the spermatogenesis process takes approximately 70 days [26, 28].

3.2 Spermatogenesis complications in mice with heat stress-induced

Globally, animal models have commonly been used to address a variety of biomedical research including, from basic science to immunology and infectious disease, oncology, and behavior or therapies. Laboratory mice are genetically heterogeneous and have been developed to be the powerhouse for biomedical research [29]. Animal models, especially the mouse model was investigated for the study of human infertility or spermatogenesis [30], effects of testes hyperthermia [10], cisplatin-induced testicular injury [31].

Healthy mice without heat stress show the complete process of spermatogenesis in which progenitor spermatogonia develop into mature spermatozoa in the seminiferous tubules (**Figure 1b**). The germinal epithelium including of cells at different stages of spermatogenic development (spermatogonia, primary spermatocytes, secondary spermatocytes, spermatids, and spermatozoa) located in invaginations or in dilations between of Sertoli cells (**Figure 1c**). Meanwhile, the heat-stress exposed mice exhibited degenerated and disorganized features of spermatogenic epithelium and reduced spermatogenic cell numbers (**Figure 1d** and **e**).

3.3 Evaluation on changes of testicular histology and spermatogenesis in mice exposed to heat stress

Johnsen's score evaluation system was used in several studies to assess testicular histology, a relatively complete scoring system with a full level of testicular histological scores given a score of 10 to 1 with a decrease in the number of cells in the lumen of the spermatogenesis [25]. This quantitative histological grading system is reliable, easily obtained and simple to prognosis for reproductive capacity in men [32, 33]. Johnsen's score is also a histopathological predictor of semen quality after a

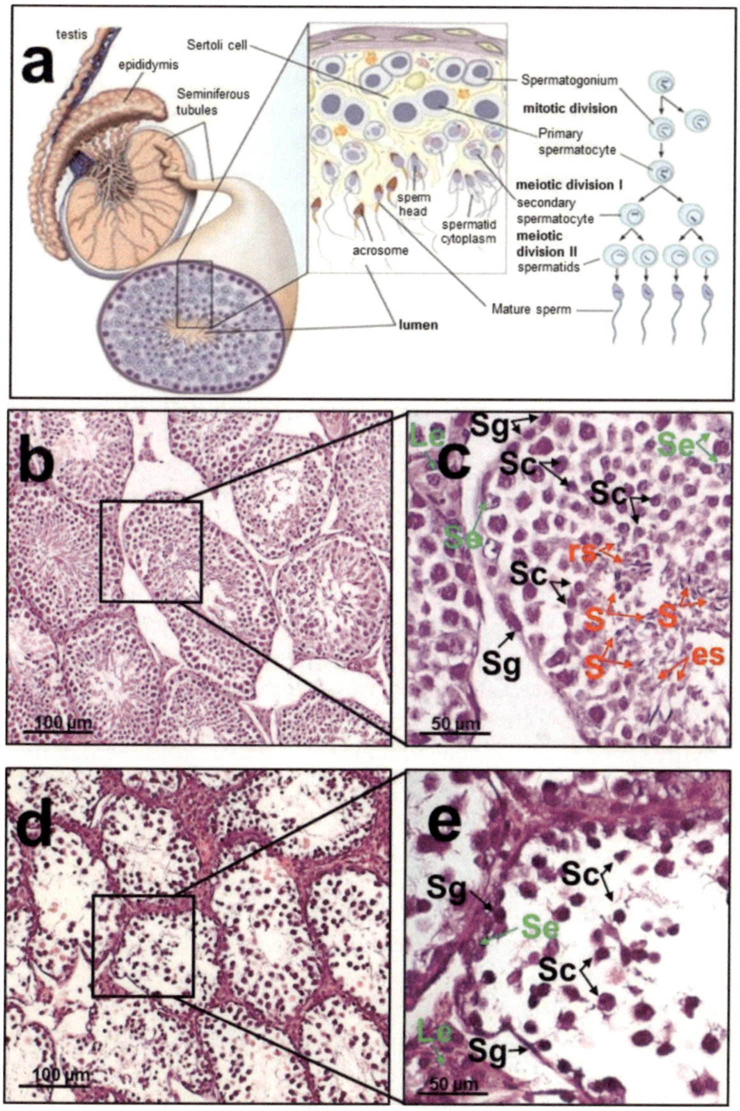

Figure 1.
Anatomy of healthy adult seminiferous tubules [26] and spermatogenesis in healthy and heat-stress exposed mice [27]. (a) Spermatogenesis takes place in the seminiferous tubules of adult testes. The magnification of a seminiferous tubule wedge highlights various cell types created in the differentiation of primordial germ cells to mature sperm. (b) Testicular structure in controlled male mice. (c) Complete spermatogenesis occurs in the seminiferous tubules. (d) Testicular structure in heat-stress exposed male mice. (e) Spermatogenesis arrest occurs in the seminiferous tubules from heat-stress exposed mice; Le: Leydig cells, Se: Sertoli cells, Sg: spermatogonia, Sc: spermatocytes, rs: round spermatids, es: elongating spermatids, S: spermatozoa.

number of treatment indications in male infertility patients to analyze and determine positive prognostic values that help improve fertility [33].

Tung et al. applied the criteria formulated by Johnsen to assess murine testicular histopathology affected by scrotal heat stress [27]. Results showed that Johnsen's score was significantly reduced in the heat stress groups compared with those in the control group, increasing the possibility of infertility in male mice. The seminiferous tubules were graded (1 to 10) according to a reduction in the number and density of germ cells from the lumen of seminiferous tubules. Standardization of Johnsen scores in the seminiferous tubule cross-sections of control and heat exposed group mice stained with hematoxylin–eosin was shown in **Figure 2**.

Johnsen scores 8, 9, and 10 indicate spermatogenesis with a few to many spermatozoa present in a section of seminiferous tubular (**Figure 2a–c**). Scores 6 and 7 indicate seminiferous tubular with no spermatozoa, but spermatids present (**Figure 2d** and **e**). Seminiferous tubular with no spermatozoa, no spermatids, but spermatocyte present is evaluated as scores 4 and 5 (**Figure 2g** and **f**). Seminiferous tubular with only spermatogonia as germ cells presence is evaluated as score 3 (**Figure 2h**), meanwhile tubular with no germ cells but Sertoli cells present is given a score 2 (**Figure 2i**). Seminiferous tubular with complete absence of cells in the seminiferous tubules is evaluated as score 1 (**Figure 2j**) [27].

3.4 The effect of heat stress on male reproduction

Genital heat stress has been known as a risk factor for male infertility. However, the exact mechanism causing impaired spermatogenesis is still unclear. Testicular histology after thermal exposure had changed such as decreased epithelial thickness, the appearance of cellular debris, fragmented cells, and absence of sperm and spermatocytes [34, 35]. The impact of prolonged heat stress negatively affects sperm quality and quantity, testicular structure with disruption of the surrounding epithelium [34]. In addition, scrotal heat stress reduces testicular weight, mitochondrial degeneration, dilatation of the smooth endoplasmic reticulum and Leydig cells lose the function of supporting stem cells [13]. Findings from this study support that the testicular structure was disorganized germinal epithelium with marked sloughing or obliteration of lumen; spermatogenesis was interrupted with the absence of many types of sperm cells in male mice after chronic scrotal heat stress exposed for 5 weeks.

The elevated testicular temperature in mammals leads to impair spermatogenesis, germ cells incur damage, decreased sperm motility, and increases the percentage of sperm with an abnormal morphology [3, 36, 37]. Paul et al. also showed the effect of heat stress on testicular function and decreased fertility in mice [38]. In addition, heat stress increases free radicals and oxidative stress resulting in apoptosis of germ cells and increased sperm DNA fracture, loss of sperm integrity [39, 40].

There are many factors that can increase testicular temperature. These factors that affect male fertility can be grouped according to habits, lifestyles [18, 19], occupational factors that must be exposed to high temperatures for long periods [20, 21], and climate change [22]. Depending on the situation, the testicles may experience transient or prolonged heat stress with varying intensity. The effects of high ambient temperature on male fertility tend to accumulate with repeated exposure over a while [7, 18].

Heat-stressed has a direct impact on work performance by increasing the risk of illness and work-related injuries. When workers are exposed to high temperatures, their bodies are unable to activate compensation regimes or recover from stressful working days that put their health in danger [41]. Boni et al. demonstrated high ambient temperature as a serious threat to reproductive function in animals and humans. Comparing parameters between the control group and the group workers exposed to heat were within the limits of normozoo sperm range, there was a serious decline in semen parameters [21]. The bakers expose to high ambient temperature have high infertility rates as shown in Al-Otaibi's study [20]. There was another large-scale epidemiological study at the Danish infertility clinic in which subjects underwent sperm examinations or infertility treatments, and obtained information on occupational exposure. It was found that the groups of workers exposed to textile dyes and lead, noise, cadmium, and mercury were all potentially infertile [42]. Similarly, Hamerezaee et al. have drawn the same results about the effects of temperature stress on semen quality when studied on workers of the steel industry

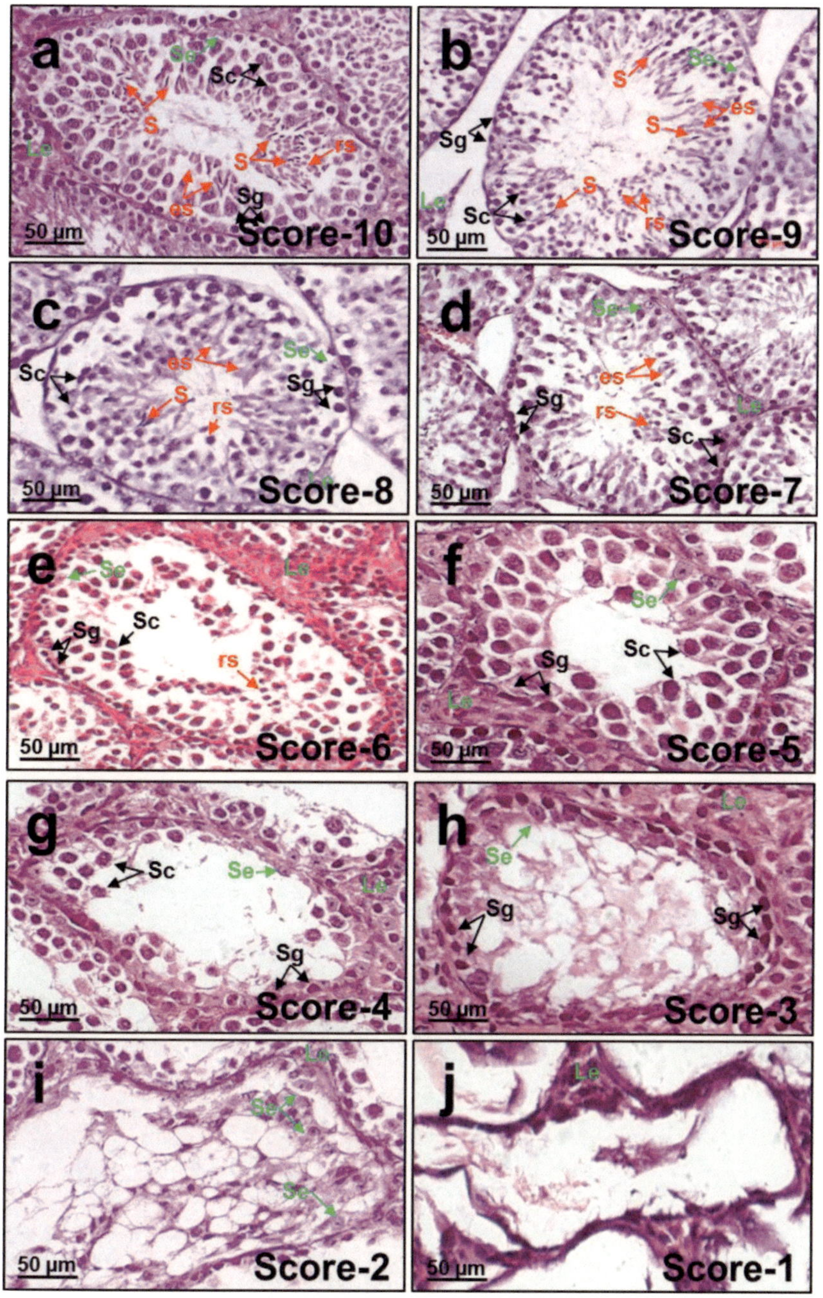

Figure 2.
Standardization of Johnsen scores in the seminiferous tubule cross-sections of control and heat exposed group mice stained with hematoxylin–eosin [27]. (a) Johnsen score 10. (b) Johnsen score 9. (c) Johnsen score 8. (d) Johnsen score 7. (e) Johnsen score 6. (f) Johnsen score 5. (g) Johnsen score 4. (h) Johnsen score 3. (i) Johnsen score 2. (j) Johnsen score 1. Le: Leydig cells, Se: Sertoli cells, Sg: spermatogonia, Sc: spermatocytes, rs: round spermatids, es: elongating spermatids, S: spermatozoa.

in Iran. The semen quality of the workers is lower than the normal limit and significantly lower than the control group [43]. Another three-year study conducted in a fertility clinic in New Orleans found that men working in buildings without air conditioning during summertime would reduce their sperm quality. Their semen samples had a significantly lower sperm concentration, number of sperms per ejaculation, and a lower percentage of sperm motility when compared those figures in summer to other seasons of the year [44]. In contrast, a study in the ceramic industry has reported that workers exposed to high temperatures had no significant difference on semen analyses except for the sperm velocity [45].

In conclusion, this chapter shows the evidence for negative effects on histopathological alterations and spermatogenesis arrest following chronic scrotal heat stress. This chapter also characterizes an animal model for studying the male reproductive system and standardized Johnsen scores system to assess murine testicular histopathology in the seminiferous tubule cross-sections.

Acknowledgements

The study was supported by the Hue university-level research projects in science and technology (DHH 2019-2104-88).

Author details

Thuan Dang-Cong[1] and Tung Nguyen-Thanh[2,3]*

1 Department of Histology, Embryology, Pathology, and Forensic, Hue University of Medicine and Pharmacy, Hue University, Hue, Vietnam

2 Faculty of Basic Science, Hue University of Medicine and Pharmacy, Hue University, Hue, Vietnam

3 Institute of Biomedicine, Hue University of Medicine and Pharmacy, Hue University, Hue, Vietnam

*Address all correspondence to: nttung@huemed-univ.edu.vn; nguyenthanhtung@hueuni.edu.vn

IntechOpen

© 2021 The Author(s). Licensee IntechOpen. This chapter is distributed under the terms of the Creative Commons Attribution License (http://creativecommons.org/licenses/by/3.0), which permits unrestricted use, distribution, and reproduction in any medium, provided the original work is properly cited.

References

[1] Boni R: **Heat stress, a serious threat to reproductive function in animals and humans**. Molecular Reproduction and Development 2019, **86**.

[2] Rockett JC, Mapp FL, Garges JB, Luft JC, Mori C, Dix DJ: **Effects of hyperthermia on spermatogenesis, apoptosis, gene expression, and fertility in adult male mice**. Biology of reproduction 2001, **65**(1):229-239.

[3] Abdelhamid MHM, Walschaerts M, Ahmad G, Mieusset R, Bujan L, Hamdi S: **Mild experimental increase in testis and epididymis temperature in men: effects on sperm morphology according to spermatogenesis stages**. Translational andrology and urology 2019, **8**(6):651-665.

[4] Shadmehr S, Fatemi Tabatabaei SR, Hosseinifar S, Tabandeh MR, Amiri A: **Attenuation of heat stress-induced spermatogenesis complications by betaine in mice**. Theriogenology 2018, **106**:117-126.

[5] Lin C, Shin DG, Park SG, Chu SB, Gwon LW, Lee JG, Yon JM, Baek IJ, Nam SY: **Curcumin dose-dependently improves spermatogenic disorders induced by scrotal heat stress in mice**. Food & function 2015, **6**(12):3770-3777.

[6] Rasooli A, Taha Jalali M, Nouri M, Mohammadian B, Barati F: **Effects of chronic heat stress on testicular structures, serum testosterone and cortisol concentrations in developing lambs**. Animal reproduction science 2010, **117**(1-2):55-59.

[7] Durairajanayagam D, Agarwal A, Ong C: **Causes, effects and molecular mechanisms of testicular heat stress**. Reproductive BioMedicine Online 2015, **30**(1):14-27.

[8] Gong Y, Guo H, Zhang Z, Zhou H, Zhao R, He B: **Heat Stress Reduces Sperm Motility via Activation of Glycogen Synthase Kinase-3α and Inhibition of Mitochondrial Protein Import**. Front Physiol 2017, **8**:718-718.

[9] Kim B, Park K, Rhee K: **Heat stress response of male germ cells**. Cellular and molecular life sciences : CMLS 2013, **70**(15):2623-2636.

[10] Kumar Roy V, Marak TR, Gurusubramanian G: **Alleviating effect of Mallotus roxburghianus in heat-induced testicular dysfunction in Wistar rats**. Pharmaceutical biology 2016, **54**(5):905-918.

[11] Zhang MH, Shi ZD, Yu JC, Zhang YP, Wang LG, Qiu Y: **Scrotal heat stress causes sperm chromatin damage and cysteinyl aspartate-spicific proteinases 3 changes in fertile men**. Journal of assisted reproduction and genetics 2015, **32**(5):747-755.

[12] Cammack KM, Antoniou E, Hearne L, Lamberson WR: **Testicular gene expression in male mice divergent for fertility after heat stress**. Theriogenology 2009, **71**(4):651-661.

[13] Durairajanayagam D, Agarwal A, Ong C: **Causes, effects and molecular mechanisms of testicular heat stress**. Reproductive biomedicine online 2015, **30**(1):14-27.

[14] Setchell B: **The effect of heat on the testes of mammals**. Anim Reprod International Symposium on Animal Biology of Reproduction 2006, **3**:81-91.

[15] Mohajeri D, Kaffashi Elahi R: **Effects of Nigella sativa on heat-induced testis damage in mouse**. Bratislavske lekarske listy 2015, **116**(4):264-269.

[16] Hamilton TRDS, Siqueira AFP, de Castro LS, Mendes CM, Delgado JdC, de Assis PM, Mesquita LP, Maiorka PC,

Nichi M, Goissis MD *et al*: **Effect of Heat Stress on Sperm DNA: Protamine Assessment in Ram Spermatozoa and Testicle**. *Oxid Med Cell Longev* 2018, **2018**:5413056-5413056.

[17] Tesarik J, Mendoza-Tesarik R, Mendoza C: **Sperm nuclear DNA damage: update on the mechanism, diagnosis and treatment**. Reproductive biomedicine online 2006, **12**(6):715-721.

[18] Durairajanayagam D: **Lifestyle causes of male infertility**. Arab J Urol 2018, **16**(1):10-20.

[19] Garolla A, Torino M, Sartini B, Cosci I, Patassini C, Carraro U, Foresta C: **Seminal and molecular evidence that sauna exposure affects human spermatogenesis**. *Human reproduction (Oxford, England)* 2013, **28**(4): 877-885.

[20] Al-Otaibi ST: **Male infertility among bakers associated with exposure to high environmental temperature at the workplace**. Journal of Taibah University Medical Sciences 2018, **13**(2):103-107.

[21] Boni R: **Heat stress, a serious threat to reproductive function in animals and humans**. Molecular Reproduction and Development 2019, **86**(10):1307-1323.

[22] Sinawat S: **The environmental impact on male fertility**. *Journal of the Medical Association of Thailand =* Chotmaihet thangphaet 2000, **83**(8):880-885.

[23] Rakesh V, Stallings JD, Reifman J: **A virtual rat for simulating environmental and exertional heat stress**. *J Appl Physiol (1985)* 2014, **117**(11):1278-1286.

[24] Paul C, Teng S, Saunders PT: **A single, mild, transient scrotal heat stress causes hypoxia and oxidative stress in mouse testes, which induces germ cell death**. Biology of reproduction 2009, **80**(5):913-919.

[25] Johnsen SG: **Testicular biopsy score count--a method for registration of spermatogenesis in human testes: normal values and results in 335 hypogonadal males**. Hormones 1970, **1**(1):2-25.

[26] Schilit S: **Uncovering Novel Cytogenetic and Molecular Etiologies for Male Infertility**. 2019.

[27] Tung NT, Phuoc DV, Thuan DC, Tam LM, Quoc Huy NV: **Assessment of testis histopathological changes and spermatogenesis in male mice exposed to chronic scrotal heat stress**. J Anim Behav Biometeorol 2020, **8**:174-180.

[28] Hunter D, Anand-Ivell R, Danner S, Ivell R: **Models of in vitro spermatogenesis**. Spermatogenesis 2012, **2**(1):32-43.

[29] Barre-Sinoussi F, Montagutelli X: **Animal models are essential to biological research: issues and perspectives**. Future science OA 2015, **1**(4):FSO63.

[30] Jamsai D, O'Bryan MK: **Mouse models in male fertility research**. Asian journal of andrology 2011, **13**(1):139-151.

[31] Whirledge SD, Garcia JM, Smith RG, Lamb DJ: **Ghrelin partially protects against cisplatin-induced male murine gonadal toxicity in a GHSR-1a-dependent manner**. Biology of reproduction 2015, **92**(3):76.

[32] Dohle GR, Elzanaty S, van Casteren NJ: **Testicular biopsy: clinical practice and interpretation**. Asian journal of andrology 2012, **14**(1):88-93.

[33] Teixeira TA, Pariz JR, Dutra RT, Saldiva PH, Costa E, Hallak J: **Cut-off values of the Johnsen score and Copenhagen index as**

histopathological prognostic factors for postoperative semen quality in selected infertile patients undergoing microsurgical correction of bilateral subclinical varicocele. Translational andrology and urology 2019, **8**(4): 346-355.

[34] Setchell BP: **The effects of heat on the testes of mammals**. *Animal Reproduction* 2006, **vol. 3**:no. 2.

[35] Gao J, Zuo Y, So KH, Yeung WS, Ng EH, Lee KF: **Electroacupuncture enhances spermatogenesis in rats after scrotal heat treatment**. Spermatogenesis 2012, **2**(1):53-62.

[36] Kanter M, Aktas C, Erboga M: **Heat stress decreases testicular germ cell proliferation and increases apoptosis in short term: an immunohistochemical and ultrastructural study**. Toxicology and industrial health 2013, **29**(2):99-113.

[37] Shiraishi K, Matsuyama H, Takihara H: **Pathophysiology of varicocele in male infertility in the era of assisted reproductive technology**. International journal of urology : official journal of the Japanese Urological Association 2012, **19**(6):538-550.

[38] Paul C, Murray AA, Spears N, Saunders PT: **A single, mild, transient scrotal heat stress causes DNA damage, subfertility and impairs formation of blastocysts in mice**. Reproduction 2008, **136**(1):73-84.

[39] Shiraishi K, Takihara H, Matsuyama H: **Elevated scrotal temperature, but not varicocele grade, reflects testicular oxidative stress-mediated apoptosis**. World journal of urology 2010, **28**(3):359-364.

[40] Nash S, Rahman MS: **Short-term heat stress impairs testicular functions in the American oyster, Crassostrea virginica: Molecular mechanisms and induction of oxidative stress and apoptosis in spermatogenic cells**. Molecular reproduction and development 2019, **86**(10):1444-1458.

[41] Momen MN, Ananian FB, Fahmy IM, Mostafa T: **Effect of high environmental temperature on semen parameters among fertile men**. *fertility and sterility* 2010, **93**(6):1884-1886.

[42] Rachootin P, Olsen J: **The risk of infertility and delayed conception associated with exposures in the Danish workplace**. Journal of Occupational and Environmental Medicine 1983, **25**(5):394-402.

[43] Hamerezaee M, Dehghan SF, Golbabaei F, Fathi A, Barzegar L, Heidarnejad N: **Assessment of Semen Quality among Workers Exposed to Heat Stress: A Cross-Sectional Study in a Steel Industry**. Safety and health at work 2018, **9**(2):232-235.

[44] Levine RJ, Bordson BL, Mathew RM, Brown MH, Stanley JM, Starr TB: **Deterioration of semen quality during summer in New Orleans**. Fertility and Sterility 1988, **49**(5):900-907.

[45] Figa-Talamanca I, Dell'Orco V, Pupi A, Dondero F, Gandini L, Lenzi A, Lombardo F, Scavalli P, Mancini G: **Fertility and semen quality of workers exposed to high temperatures in the ceramics industry**. Reproductive Toxicology 1992, **6**(6):517-523.

Section 3

Clinical Management and Epigenetics

Chapter 7

Methods of Sperm Selection for In-Vitro Fertilization

Abimibola Nanna

Abstract

50–60% of infertility cases are as a result of male infertility and infertile men semen sample is characterize with poor motility, abnormal morphology, low sperm concentration, azoospermic and increased levels of sperm DNA damage. As a result of this heterogeneity of the ejaculate, sperm selection has become a necessary step to carry out prior to in vitro fertilization. Furthermore, the choice of sperm cell selection techniques depend on sperm concentration and sperm biology and the recovery of highly functional sperm cell population depend on the combination of more than one technique in some cases. The regular sperm cell selection methods in ART laboratory are swim up, density gradient, simple wash and other advanced and emerging sperm selection techniques which include hyaluronic acid mediated sperm binding, Zeta potential, hypoosmotic swelling test, magnetic activated cell sorting and microfluidic separation of sperm cells. The various methods have its own advantages and disadvantages which may be applicable to the individual need of infertile men and its effect on ART outcome.

Keywords: male infertility, sperm selection techniques, semen

1. Introduction

Male infertility accounts for 50–60% of infertility cases and abnormal semen qualities like low motility, low sperm concentration, abnormal morphology and increased levels of sperm DNA damage are characteristic of infertile men sample [1]. Furthermore high level of reactive oxygen species (ROS) are found in 40–88% of sperm samples of infertile men [2] and physiological sperm functions such as capacitation, acrosome reaction and hyperactivation requires low ROS concentration while ROS overproduction is usually due to the inability of antioxidant to neutralize ROS [3]. Also, decrease sperm motility, DNA integrity and viability, increase midpiece defects are cause by oxidative stress from high level of ROS and decreased levels of antioxidant [3]. In addition, lower in vitro fertilization pregnancy rate, irregular preimplantation development, early loss of pregnancy and decreased rate in ART conceived offsprings are correlated with poor DNA integrity [4].

As a result of the above heterogeneity of ejaculate (understanding of sperm physiology) and male gamete integrity rule in both fertilization and embryogenesis, has led to an increased demand on sperm selection techniques. Sperm biology, sperm concentration, volume and life time invitro are the fundamental challenge of sperm selection and sperm selection process ideal time is about 10 minutes for 1 ml of sperm sample containing 100 million/ml. This shows a very high biological sorting rate of −100 KHZ and the current cell sorting technologies has lower value of this [5].

There are various sperm selection techniques in Assisted Reproductive Technology (ART) and these techniques try to replicate vivo, the natural process in which quality sperms are selected from other constituents of the ejaculate as they actively move through the cervical mucus [6]. The simple wash, swim up and density gradient are the three most common sperm selective methods for sperm preparation in ART and other additional methods that will be discussed in this chapter.

2. Sperm production

At ejaculation, semen consists of a suspension of spermatozoa which is stored in the epididymis and is mixed with the secretion of the accessory glands. The prostate and the seminal vesicles are the main glands while the bulbourethral glands and the epididymis are responsible for the minor contribution of the ejaculate. The seminal fluid consists of prostatic fraction which is rich in sperm cells and the vesicular fraction which is less in spermatozoa. Furthermore, it is essential to have a complete sample volume collection and not to lose the first rich sperm fraction during masturbation which might make semen analysis difficult. Thus complete semen sample production is the first step throughout sperm preparation [7].

In additional, sample production is carry out through masturbation with sterile specimen container. This is done after abstinences of 2–3 days which maximize conception [8]. Furthermore, patient should be encourage onsite production of semen sample which avoids extreme temperature exposure while offsite semen production should avoid spermicidal effect from lubricant and samples should get to the andrology/in vitro fertilization (IVF) laboratory within 30–40 minutes without extreme environmental temperature exposure [9].

3. Choice of sperm selection techniques

The selection or preparation techniques use in sperm separation in IVF laboratory depends on the characteristic of the semen sample when subjected to semen analysis [10]. Furthermore, sperm cells should be separated from seminal plasma as early as possible and in vitro fertilization capacity diminishes permanently when sperm cells are not separated from seminal plasma within 30 minutes of ejaculation [11]. Furthermore, World Health Organization recommend sperm cell separation from seminal plasma within one hour of ejaculation and this will limit damage from leucocytes and other cells present in the semen [12]. Semen samples characterize with severe oligo and athenozoospermia are separated using simple wash and normazoospermia sample are separated with swim up or density gradient. Also, suboptimal quality semen sample are separated by density gradient [13].

4. Steps to maximize quality of sperm during sperm selection techniques

A. Bring gradient and sperm wash media to room temperature before use and this will protect sperm cells from cold shock. Also visual detection of contamination on the media will be seen when condensation on the media bottle disappear.

B. Media content should not be use when it is cloudy or hazy.

C. Use individual pipette for each bottle of media.

D. The PH of media is alter with prolong exposure to a 5% co2 environment which will affect their nature and performance.

E. For highly viscous semen, dissolve 5 mg of trypsin in 1 ml of sperm washing media and add it to the highly viscous semen for about 5 minutes before loading the upper gradient. This will allow increase yield of motile sperm without causing any great damage to the motile sperm.

F. Avoid excessive loading of semen on the gradient as this will cause a rafting phenomena. Rafting is define as the present of aggregate of desirable as well as undesirable components of the semen in post centrifuged pellet. For example, a gradient of 1-2 ml of upper and lower gradient, semen volume of 1 ml should be added accordingly.

G. Gradient should be use within one hour of creation. Delay usage of the gradient could lead to the two phase mixing with each other [14].

5. Simple wash method

It involved one or two centrifugation of semen sample in order to separate the sperm pellet from the seminal plasma. This process does not significantly decrease sperm count, normal morphology, and motility remain unchanged but there is increase in rapid forward progression and hypermotility of sperm cells in the post washed sample. Furthermore, this process is use for cases of severe oligospermia, asthenozoospermic semen sample and it is recommended method for insemination-ICSI and not for standard insemination procedure in IVF. Also, centrifugation process in this method causes additional harm to sperm cells by the production of reactive oxygen species (ROS) by leucocytes and abnormal sperm cells. ROS production causes DNA damage in spermatozoa, decreased sperm motility, increase number of apoptotic sperm cells and decrease sperm plasma membrane integrity (**Figure 1**).

5.1 Method

1. Semen sample is mix well.

2. Add supplement media to semen sample in a ratio of 1:1 in 15 ml conical tube.

3. Centrifugation is done at 1800 g for 5–10 minutes.

4. Aspirate the supernatant.

5. Add 0.1–0.5 ml of the culture media to the pellet for swim up.

6. Insemination can be done and sperm concentration and motility can be determined using WHO 2010 protocol [15, 16].

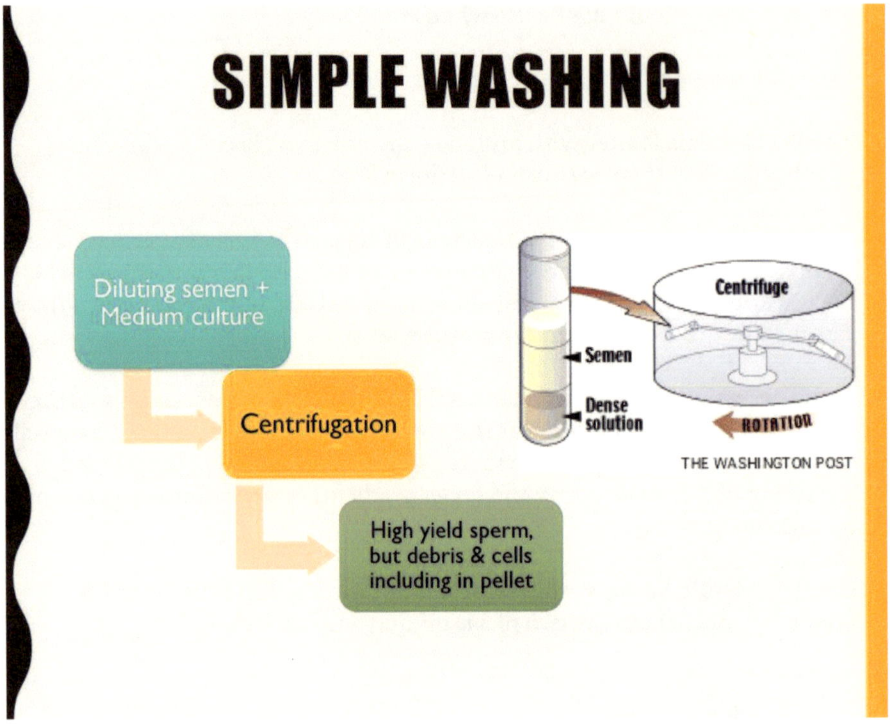

Figure 1.
Above indicate a typical simple wash procedure.

6. Swim up from semen

It is recommended for normazoospermic semen quality with high sperm count and good motility. Furthermore, it is carry out in round bottom tube place at an angle of 45 degree and this method favors the selection of motile sperm with intact membrane which result in higher clinical pregnancy rate in the IVF laboratory (**Figure 2**).

6.1 Method

1. Incubate 4–5 falcon sterile round bottom tube at 37 degree Celsius for few hours and add 1–2 ml of sperm wash medium.

2. Allow semen to liquefy and carry out semen analysis using WHO protocol.

3. Gently underlay liquefied semen (1–2 ml) in the bottom of the sterile tube.

4. Place the round bottom tube at 45 degree position in a humidified incubator at 37 degree or at room temperature for 30–60 minutes with tubes tightly capped. The motile sperm cells migrate upward into the sperm wash medium.

5. Aspirate the upper and middle section of the medium in the tubes and combined it in a conical centrifuge tube.

6. Centrifuge the pooled swim up fractions at 300 g for 10 minutes and remove supernatant and repeat centrifugation with additional 2 ml sperm wash.

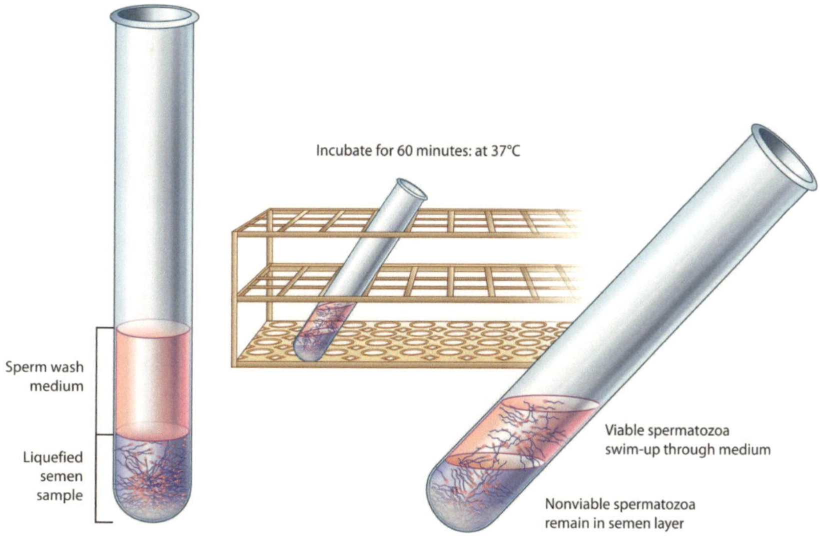

Figure 2.
Swim up from semen where the round bottom tube is place at an angle of 45 degree for 30–60 minutes. Active sperm cells swim to the top and are aspirated.

7. Remove the supernatant.

8. Suspend the pellet in equilibrated bicarbonate based IVF medium (0.1–0.5 ml) and incubate at 37 degree in 5–6% co2 environment for 30 minutes.

9. Post wash analysis is carry out for sperm concentration and motility.

10. Use for IVF or ICSI insemination [15, 17].

7. Density gradient centrifugation

It separates spermatozoa based on their density and at the end of each centrifugation, each spermatozoon is found at the gradient level that corresponds to its density. Furthermore, normal morphological spermatozoon has at least 1.10 g/ml density while abnormal one has density between 1.06–1.09 g/ml [18].

In addition, following centrifugation, the leucocytes, cell debris and sperm cell with abnormal morphology with poor motility are found at the interphase between seminal plasma and 45%, 45% and 90%. Also, the pellet at the bottom of the tube is characterized with highly motile, morphologically normal, viable spermatozoa [19]. Furthermore, if the volume of each gradient is lower than 1 ml, greater number of motile spermatozoa can be recovered, as a result of the spermatozoa have to migrate for a less distance between layers [20]. This method is used for normazoospermic semen and suboptimal semen qualities and it is recommended method for insemination IVF or ICSI [15].

7.1 Gradient preparation

1. Add 1-2 ml of lower phase (90%) at the bottom of the 15 ml conical tube.

2. Add 1–2 ml of upper phase (45%) on top without mixing the two gradients.

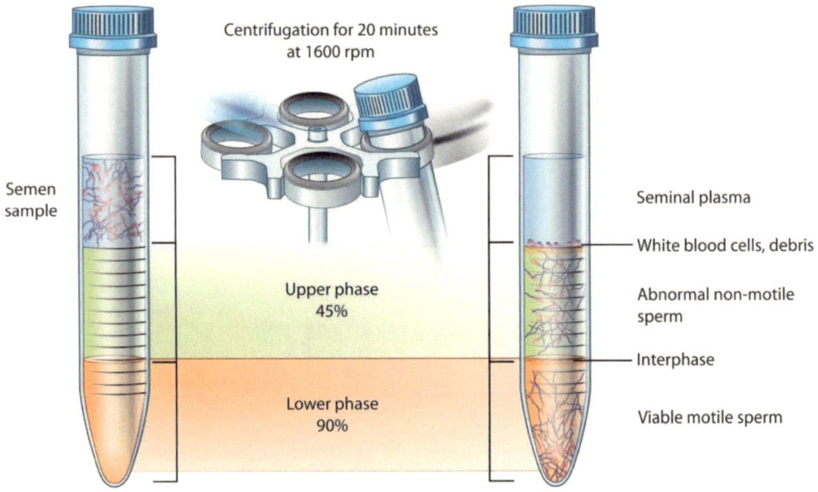

Figure 3.
Indicate sperm cell separation of density gradient.

3. Add 1-2 ml of liquefied semen on top and centrifuged for 15–20 minutes at 1600–3000 rpm.

4. Remove supernatant without disturbing the pellet at the bottom of the tube.

5. Aspirate the pellet by adding 0.1mlof sperm wash and transfer it to fresh conical tube containing 2–5 ml of sperm wash media.

6. Centrifuge at 1600–3000 rpm for 10 minutes, discard the supernatant and repeat this step.

7. Re-suspend the pellet in 0.1–0.5 ml of culture media (Fertilization media) and to allow swim up for 15–20 minutes in the incubator.

8. Carry out post wash analysis for sperm concentration and motility and inseminate for IVF (100,000 progressive motile sperm/ml) and ICSI [15].

One of the advantage of this method is that, it eliminates the majority of leucocytes in the ejaculate, easy to carry under sterile condition, it consume less time and major disadvantage is that spermatozoa recovered from this method has lower DNA integrity as compare to spermatozoa from swim up method (**Figure 3**) [21].

8. Sperm preparation for testicular sperm, epididymal sperm and retrograde ejaculation

Sperm Cells can be gotten from cases of epididymal obstruction or complete azoospermia through their epididymis or the testicular tissue. These sperm cells gotten from the process can be separated through Simple wash technique if the number of sperm cells isolated is low while density gradient is used when the number of sperm cells collected is enough.

8.1 Epididymal sperm preparation

1. Examine the epididymal aspirate when pour into the culture dish through the dissected microscope.

2. From your view, determine if simple wash method or density gradient will be use.

3. Pour the aspirate into the conical tube and add 1–2 ml of sperm wash media.

4. Centrifuge the conical tube at 400 g for 10 minutes and discard the supernatant or use density gradient base on the number of sperm cells.

5. Wash with RBC lysis buffer if the pellet is mix with blood cells.

6. Add 2 ml of RBC lysis buffer and gently mix, centrifuge at 400 g for 5 minutes.

7. Aspirate the supernatant without affecting the pellet.

8. Add 1–2 ml of sperm wash to the pellet and centrifuge at 400 g for 10 minutes.

9. Pellet is suspended in 0.5 ml of sperm wash, for ICSI purpose.

8.2 Testicular sperm preparation

1. Add 1-2 ml of culture medium to the testicular tissue to wash away the red blood cells.

2. Transfer the tissue to Petri dish and add another few drops of culture medium and this is to keep it moist.

3. Add additional 1 ml of culture medium and minced with sterile scissors.

4. Aspirate the suspension and pass it through 21 gauze needle which is attached to a 3 ml syringe. The passage is done 2–3 times and the process, sperm cells are released from the seminiferous tubules in the small pieces and also dislodges the sperm from the lumen.

5. Transfer the suspension into a conical tube and allow standing at room temperature for 5 minutes. Tissue clumps and seminiferous tubules will settle at the bottom of the tube.

6. Supernatant is transfer to another conical tube and centrifuged at 300 g for 10 minutes.

7. Supernatant is discarded.

8. If there are red blood cells with the pellet, wash with 2 ml of RBC lysis buffer by centrifuging at 300 g for 5 minutes.

9. Remove supernatant and add 0.5-1 ml of culture media to the pellet and it ready for ICSI [22, 23].

8.3 Retrograde ejaculation sample preparation

1. Patients abstain for 3 days.

2. The evening before the day of sample production, patient drink 250 ml of alkaline drink (Bicarbonate of soda).

3. The morning of production, patient empty his bladder.

4. Patient takes another 250 ml of alkaline drink.

5. One hour later, when patient has the urge to pass urine, 3–4 sterile container containing 20 ml of sperm medium each is given to the patient to pass urine and bring to the IVF laboratory.

6. All tubes are centrifuge together at 400 g for 15 minutes.

7. Supernatant is discarded and pellet is re-washing again with 5 ml of sperm medium at 400 g for 10 minutes.

8. Supernatant is discarded and 0.5–1 ml of sperm medium is added to the pellet for swim up.

9. Wash sperm ready for post wash sperm concentration and motility assessment, ICSI procedure [15, 23].

9. Emerging sperm selection techniques before ICSI procedure

There are various advanced techniques of sperm preparation which is based on spermatozoa surface charge and morphology and this has overcome the limitation of classical sperm selection procedures. This techniques include Zeta Potential, Hypoosmotic Swelling Test (HOST), Hyaluronic acid-mediated sperm selection, magnetic-activated cell sorting (MACS) and the microfluidics which is the latest.

10. Zeta potential

It is the electrical charge potentials that exist between the sperm surrounding and negatively charge sperm membrane. The epididymal protein present in the surface of sperm membrane is responsible for the negatively charge on this sperm membrane [24] and this is lower in sperm cells with damage DNA [25]. Furthermore, the mature sperm has zeta potentials between −16 to −120 mV [26].

10.1 Method

1. Pipette 100 ul of wash sperm into 15 ml conical tube which is suspended in serum free HEFES-HIF medium.

2. Rotate or rub the tube on a latex glove a couple of times. This will make a negative charge sperm to stick to the wall of a positively charge plastic tube.

3. Invert the tube and immature abnormal sperm cells are discarded.

4. Tube is place on the rack for 1 minute at room temperature without agitation, then centrifuge at 300 g for 5 minutes.

5. Suspend the pellet in 0.2 ml of serum supplement medium, examine under the microscope after swim up and use for ICSI [25].

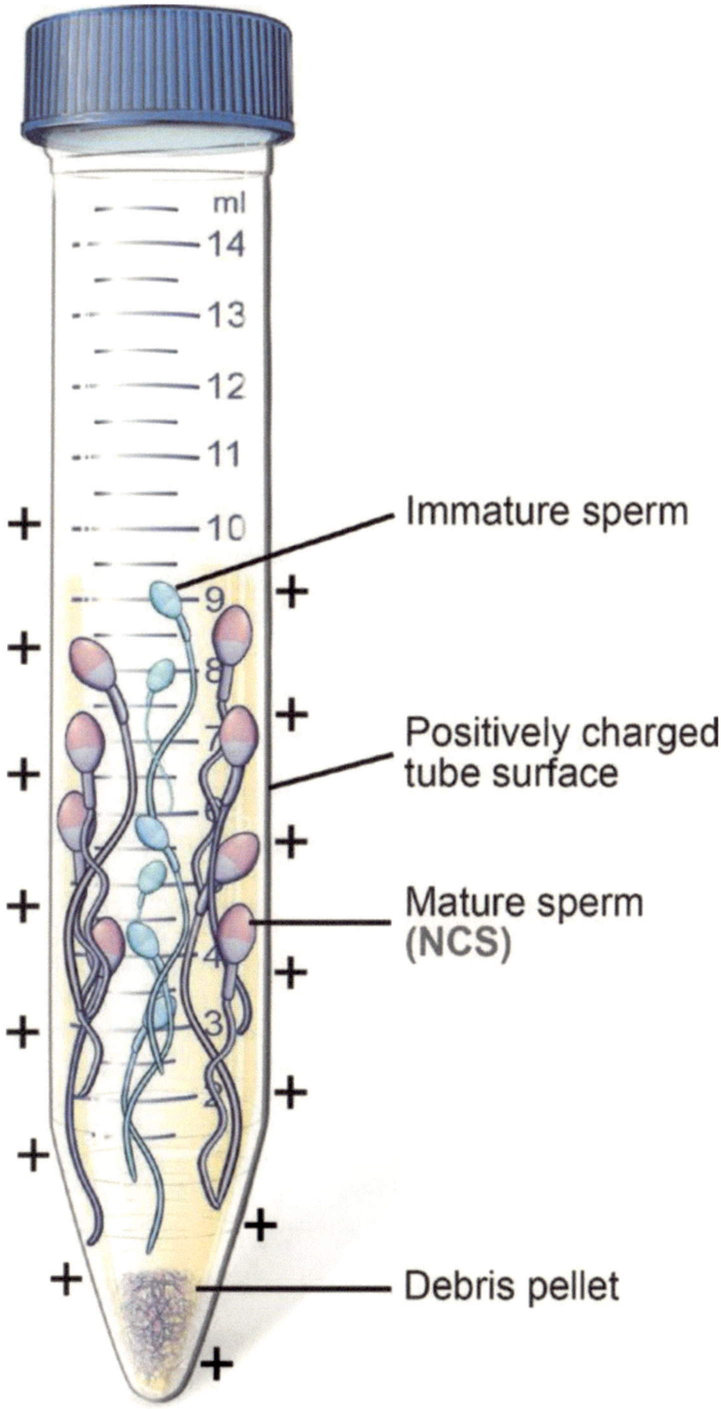

Figure 4.
Above show zeta potential process during the separation of mature and immature spermatozoa.

Also, the method is inexpensive, easy to carry out. Sperm with intact DNA, superior motility and normal morphology can be separated with this method and negative zeta potentials sperm in IVF had a higher fertilization (65.79%) compared with sperm selected with double density gradient centrifugation. Furthermore, there is increase fertilization rate and possible pregnancy rate in infertile couples with male factor infertility whose semen sample was selected with double density gradient/zeta potential as compared to double density gradient alone during an ICSI procedure in a randomized prospective study (**Figure 4**) [27].

11. Hyaluronic acid-mediated sperm binding (HA)

The plasma membrane of acrosome-intact sperm cell has hyaluronic acid receptors (HA) which are indicative of sperm maturity. Also, the extracellular matrix of the cumulus oophorus contains hyaluronic receptors as its main component. Mature sperm cell bind with hyaluronic acid in the oocytes, therefore having a better chance of oocyte fertilization.

Physiological intracytoplasmic sperm injection (PICSI) plastic dish has spots containing HA which is attached to its base or a viscous medium containing HA (Sperm slow). 2ul of sperm suspension is added to the spots and incubated under oil for 5 minutes. Mature sperm will attached its head to HA and the sperm can easily be selected by ICSI injecting pipette for ICSI procedure [28].

Furthermore, the selection of sperm with reduced oxidation stress gotten with the use of HA binding which excludes immature sperm with cytoplasmic extrusion, presence of sperms with histones and DNA fragmented sperm cells. Also, significant increased fertilization rate of oocyte injected with HA-selected sperm has been reported in one study [29] while better embryos quality has also been reported in another study [30].

12. Magnetic activated cell sorting (MACS)

It uses a strong magnetic field principle where non-apoptotic and apoptotic sperm cells are pass through this field. It is only the non-apoptotic sperm cell that is retain in the field and separated from the apoptotic ones. Furthermore, phosphatidylserine is a phospholipid found on the plasma membrane of spermatozoa and it moves to the outer surface when the plasma membrane is damaged (apoptotic sperm cell) and this aid selection of non-apoptotic spermatozoa in ART [31].

The combination of density gradient and MACS has given higher quality sperm than density gradient alone. MACS removes already damaged spermatozoa with altered membrane, activated apoptosis signaling and DNA fragmentation while density gradient removes immature sperm cells, debris and leucocytes. Furthermore, the use of MACS techniques allows the selection of sperm cells with improved morphology, motility, viability and greatly improved fertilization rates [32]. Thus, increase cleavage rate and clinical pregnancy rate has been reported for sperm cell used in ICSI selected by MACS compared with density gradient in oligo-astheno- and teratozoospermic men [33].

13. Microfluidic separation of sperm cells

This is the latest sperm selection techniques and the device is made up of microchannels made from polydimethylgiloxane (PDMS) silicon polymers which are nontoxic and transparent [34].

The boundary following behavior is the principle use in selection of spermatozoa in the popular passive microfluidic device. It consists of a radial network of channels (52u width) which is involved in the separation of spermatozoa into left, right and straight swimmers. 200 ul of raw semen is loaded into the inner ring with plastic syringe and kept undisturbed for 15 minutes at 37 degree Celsius. Motile sperm cells are collected at the outlet microchannels while dead or immotile sperm cells are retained in the inlet.

Furthermore, small volume of sperm is required for this process and DNA fragmentation is significantly decreased in sperm separated with this method (Sperm damage in swim up is 16.4 and and microfluidics is 8.4%). Furthermore, the use of microfluidics device in sperm selected for ICSI treatment in porcine, yielded short time for ICSI treatment and increased the number of viable embryos without the reduction of the invitro production efficiency [35].

14. Hypoosmotic swelling test (HOS test)

Jeyendran et al. developed hypoosmotic swelling test to evaluate the functional integrity of the sperm membrane [36]. Tail swelling and curving is exhibited by viable sperm with normal membrane function when exposed to hypoosmotic condition [37]. Furthermore, its application seems to be a promising method to identify live spermatozoa for ICSI and study conducted showed that injected spermatozoa selected by HOS test gave higher fertilization rate than randomly selected spermatozoa [38].

In another study, 50% sperm media and 50% of Millipore-grade water was used to prepare HOS test microdroplet. Immotile spermatozoa was introduced into the microdroplet and after 10 seconds, the viable spermatozoa whose tails were curved and swollen, were further selected, transferred to hepes-buffered microdroplet and wash three times is sperm medium for re-equilibration before been transferred to PVP for ICSI purpose. According to the group, improved fertilization and pregnancy rates were recorded as compared with non use of HOS test (**Figure 5**) [39].

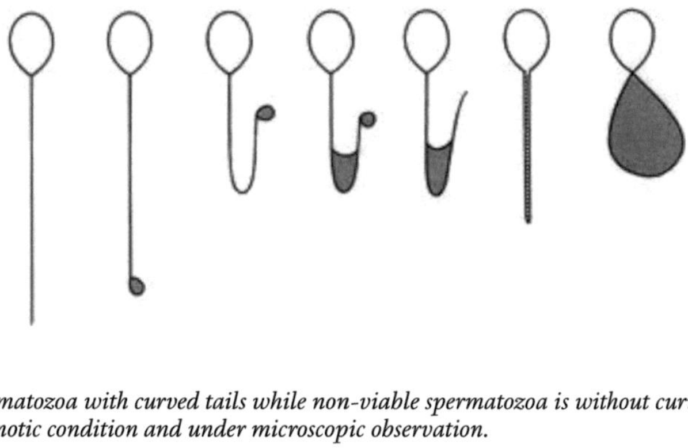

Figure 5.
Indicate viable spermatozoa with curved tails while non-viable spermatozoa is without curved tail when subjected to hypoosmotic condition and under microscopic observation.

15. Conclusion

Sperm selection techniques are very important and the choice of technique depends on semen quality of each male partner. The appropriate selection of sperm cells with lower DNA damage will improve the chance of fertilization rate, cleavage rate, Implantation and pregnancy rate in IVF laboratory.

Summary of the advantages and disadvantages of each method.

Methods	Advantage	Disadvantage
Simple wash	It is simple to perform and produce high yield of spermatozoa.	Increase DNA damage of spermatozoa.
Density gradient	It is easy to perform under sterile condition in a short time.	Contamination with endotoxins a risk factor.
	It eliminates majority of leucocytes in the ejaculate.	Sperm cells recovered with this method has lower DNA integrity.
Swim up	Spermatozoa with high DNA integrity are recovered.	It consumes more time of preparation.
	It is simple and not expensive.	
MACS	Molecular level is involved.	It is use in conjunction with density gradient to remove substance like leucocytes, seminal plasma and apoptotic spermatozoa.
	It is non-invasive, convenient and rapid.	
	It differentiate and separate non-apoptotic spermatozoa from apoptotic spermatozoa.	
HA	It supports sperm cells selection with good DNA integrity in severe oligozoospermic sample.	It is done in conjunction with intracytoplasmic sperm injection (ICSI) and simple wash method.
HOS test	Identification of live spermatozoa in severe asthenozoospermic sample.	Additional time needed to perform this method before ICSI.
	It is use for normal semen sample for identification of intact membrane sperm cell before ICSI.	
	It simple to perform.	
Zeta potential	It is not costly and easy to carry out.	It operates in conjunction with density gradient procedure.
	Recovery of spermatozoa with normal morphology, superior motility and intact DNA.	
Microfluidic device	It uses small volume of semen volume.	Not yet available done in human IVF treatment but available in porcine spermatozoa.
	Spermatozoa with significant decrease of DNA fragmentation are recovered.	

Author details

Abimibola Nanna[1,2,3,4,5]

1 University of Benin, Benin City, Nigeria

2 University of Leeds, Leeds, United Kingdom

3 Delta State University Teaching Hospital, Oghara, Nigeria

4 Shepherd Hospital and Fertility Centre, Warri, Nigeria

5 St. Jude's Fertility Institute, Abuja, Nigeria

*Address all correspondence to: bolanannal@gmail.com

IntechOpen

© 2021 The Author(s). Licensee IntechOpen. This chapter is distributed under the terms of the Creative Commons Attribution License (http://creativecommons.org/licenses/by/3.0), which permits unrestricted use, distribution, and reproduction in any medium, provided the original work is properly cited.

References

[1] Agarwal A, Mulgund A, Hamada A, Chyatte MR. A unique view on male infertility around the globe. Reprod Biol Endocrinol. 2015;13(1):37.

[2] Agarwal A, Sharma RK, Nallella KP, Thomas AJ Jr, Alvarez JG, Sikka SC. Reactive oxygen species as an independent marker of male factor infertility. Fertil Steril. 2006; 86(4):878-885.

[3] Wright C, Milne S, Leeson H. Sperm DNA damage caused by oxidative stress: Modifiable clinical, lifestyle and nutritional factors in male infertility. Reprod Biomed Online. 2014; 28(6):684-703.

[4] Agarwal A, Cho C-L, Majzoub A, Esteves SC. The Society for Translational Medicine: Clinical practice guidelines for sperm DNA fragmentation testing in male infertility. Transl Androl Urol. 2017;6(Suppl 4): S720.

[5] Nosrati R, Vollmer M, Eamer L, San Gabriel MC, Zeidan K, Zini A, et al. Rapid selection of sperm with high DNA integrity. Lab Chip. 2014;14(6): 1142-1150.

[6] Franken DR, Claasens OE, Henkel RR. Sperm preparation techniques, X/Y chromosome separation. In: Acosta AA, Kruger TF (Eds). Human Spermatozoa in Assisted Reproduction. 2nd edn. USA: Informa Healthcare. 1996. p. 277-294.

[7] Bjorndahl, L. and Kvist, U. (2003). Sequence of ejaculation affects the spermatozoon as a carrier and its message. Reprod Biomed Online, Vol. 7, No. 4 (October/November 2003), pp. 440-8.

[8] Marshburn, P.B., Alanis, M., Matthews, M.L., Usadi, R., Papadakis, M.H., Kullstam, S. and Hurst, B.S. (2010). A short period of ejaculatory abstinence before intrauterine insemination is associated with higher pregnancy rates. Fertil Steril, Vol. 93, No. 1 (January 2010), pp. 286-288.

[9] Allamaneni SSR, Agarwal A, Rama S, Ranganathan P, Sharma RK. Comparative study on density gradients and swim-up preparation techniques utilizing neat and cryopreserved spermatozoa. Asian J Androl. 2005;7: 86-92.

[10] Canale, D., Giorgi, P.M., Gasperini, M., Pucci, E., Barletta, D., Gasperi, M. and Martino, E. (1994). Inter and intra-individual variability of sperm morphology after selection with three different techniques: Layering, swim up from pellet and percoll. J Endocrinol Invest, Vol. 17, No. 9 (October 1994), pp. 729-732.

[11] Bjorndahl L, Mortimer D, Barratt CLR, Castilla JA, Menkveld R, Kvist U, et al. Sperm Preparation. A Practical Guide to Basic Laboratory Andrology. 1st edn. USA: Cambridge University Press 2010. pp. 167-187.

[12] Sperm Preparation Techniques. In: Cooper TG, Aitken J, Auger J, Baker GHW, Barratt CLR, Behre HM, et al. (Eds). World Health Organization Laboratory Manual for the Examination and Processing of Human Semen. 5th edn. Switzerland: WHO Press 2010. pp. 161-168.

[13] Methods for Sperm Selection for In Vitro Fertilization, Nicolas M. and Pablo Bosch, 2012.

[14] Henkel RR, Schill WB. Sperm preparation for ART. Reprod Biol Endocrinol 2003; 1:108.

[15] Kay Elder and Brian Dale. (2011) In-vitro fertilization. 3rd Edition. Cambridge University Press, United Kingdom.

[16] Morrell, J. M., Garcia, B. M., Pena, F. J. and Johannisson, A. Processing stored stallion semen doses by single layer centrifugation. Theriogenology, 2011 Vol. 76, No. 8, pp. 1424-32, ISSN 1879-3231 (Electronic) 0093-691X (Linking).

[17] Mortimer, D. Sperm preparation methods. J Androl, 2000 Vol. 21, No. 3, pp. 357-66, ISSN0196-3635 (Print) 0196-3635 (Linking).

[18] Oshio S, Kaneko S, Iizuka R, Mohri H. Effects of gradient centrifugation on human sperm. Arch Androl 1987; 19(1):85-93.

[19] Bourne H, Edgar DH, Baker HWG. Sperm preparation techniques. In: Gardner DK, Weissman A, Howles CM, Shoham Z (Eds). Textbook of Assisted Reproductive Techniques: Laboratory and Clinical Perspectives. 2nd edn. USA: Informa Healthcare 2004. pp. 79-91.

[20] Ilaria Natali (2011). Sperm Preparation Techniques for Artificial Insemination – Comparison of Sperm Washing, Swim Up, and Density Gradient Centrifugation Methods, Artificial Insemination in Farm Animals, Dr. Milad Manafi (Ed.), ISBN: 978-953-307-312-5, InTech, Available from: http://www.intechopen.com/books/artificialinsemination-in-farm-animals/sperm-preparation-techniques-for-artificial-insemination-comparison-of-spermwashing-swim-up-and-den.

[21] Zini A, Finelli A, Phang D, Jarvi K. Influence of semen processing technique on human sperm DNA integrity. Urology. 2000; 56(6):1081-4. 33.

[22] Bhushan K. Gangrade. Cryopreservation of testicular and epididymal sperm techniques and clinical outcomes of assisted conception. Clinics 2013;68 (51):131-140.

[23] Schoysman R, Vanderzwalmen P, Nijs M, Segal L, Segal-Bertin G, Geerts L, et al. Pregnancy after fertilization with human testicular spermatozoa. Lancet. 1993; 342:1237.

[24] Ishijima SA, Okuno M, Mohri H. Zeta potential of human X- and Y-bearing sperm. Int J Androl. 1991; 14(5):340-347.

[25] Chan PJ, Jacobson JD, Corselli JU, Patton WC. A simple zeta method for sperm selection based on membrane charge. Fertil Steril. 2006;85(2): 481-486.

[26] Ishijima, S. A., Okuno, M. and Mohri, H. Zeta potential of human X- and Y-bearing sperm. Int J Androl, 1991 Vol. 14, No. 5, pp. 340-7, ISSN 0105-6263 (Print) 0105-6263 (Linking).

[27] Polak de Fried E, Denaday F. Single and twin ongoing pregnancies in two cases of previous ART failure after ICSI performed with sperm sorted using annexin V microbeads. Fertil Steril. 2010;94(1):351.e15-8.

[28] Huszar G, Jakab A, Sakkas D, Ozenci C-C, Cayli S, Delpiano E, et al. Fertility testing and ICSI sperm selection by hyaluronic acid binding: Clinical and genetic aspects. Reprod Biomed Online. 2007;14(5): 650-663.

[29] Nasr-Esfahani, M. H., Razavi, S., Vahdati, A. A., Fathi, F. and Tavalaee, M. (2008). Evaluation of sperm selection procedure based on hyaluronic acid binding ability on ICSI outcome. J Assist Reprod Genet, Vol. 25, No. 5, pp. 197-203, ISSN 1058-0468 (Print) 1058-0468 (Linking).

[30] Parmegiani, L., Cognigni, G. E., Bernardi, S., Troilo, E., Ciampaglia, W. and Filicori, M. (2010a). "Physiologic ICSI": Hyaluronic acid (HA) favors selection of spermatozoa without DNA fragmentation and with normal nucleus, resulting in improvement of embryo quality. Fertil Steril, Vol. 93, No. 2, pp.

598-604, ISSN 1556-5653 (Electronic) 0015-0282 (Linking).

[31] Grunewald S, Paasch U, Glander H-J. Enrichment of non–apoptotic human spermatozoa after cryopreservation by immunomagnetic cell sorting. Cell Tissue Bank. 2001;2(3):127-133.

[32] Said TM, Grunewald S, Paasch U, Glander H-J, Baumann T, Kriegel C, et al. Advantage of combining magnetic cell separation with sperm preparation techniques. Reprod Biomed Online. 2005;10(6):740-746.

[33] Dirican EK, Özgün OD, Akarsu S, Akın KO, Ercan Ö, Uğurlu M, et al. Clinical outcome of magnetic activated cell sorting of non-apoptotic spermatozoa before density gradient centrifugation for assisted reproduction. J Assist Reprod Genet. 2008;25(8): 375-81. 44.

[34] Tasoglu S, Safaee H, Zhang X, Kingsley JL, Catalano PN, Gurkan UA, et al. Exhaustion of racing sperm in nature-mimicking microfluidic channels during sorting. Small. 2013; 9(20): 3374-3384.

[35] Eamer L, Vollmer M, Nosrati R, San Gabriel MC, Zeidan K, Zini A, et al. Turning the corner in fertility: High DNA integrity of boundary-following sperm. Lab Chip. 2016; 16(13): 2418-2422.

[36] Jeyendran RS, Van der Ven HH, Perez-Pelaez M, Crabo BG, Zaneveld LJ. Development of an assay to assess the functional integrity of the human sperm membrane and its relationship to other semen characteristics. J Reprod Fertil. 1984; 70(1):219-228, DOI:10.1530/jrf.0.0700219

[37] Desmet B, Joris H, Nagy Z, Liu J, Bocken G, Vankelecom A, et al. Selection of vital immotile spermatozoa for intracytoplasmic injection by the hyposmotic swelling test. Hum Reprod. 1994;9 (suppl 4):24-31.

[38] Casper RF, Cowan L, Meriano JS, Lucato ML, Jarvi KA. The hypoosmotic swelling test for selection of viable sperm for intracytoplasmic sperm injection in men with complete asthenozoospermia. Fertil Steril. 1996;65(5):972-976.

[39] Sallam HN, Farrag A, Agameya A, El-Garem Y, Ezzeldin F, Sallam A. Using the modified hypo-osmotic swelling test for the selection of immotile testicular spermatozoa in ICSI – A randomized controlled study. Fertil Steril. 2005;84:S373-S3S4, DOI:10.1016/j.fertnstert.2005.07.977

Chapter 8

Management of Post-Circumcision Glans/Penile Necrosis

Yusuf Arikan and Ali Ayten

Abstract

Circumcision is the surgical removal of the skin covering the glans and is one of the oldest and most common surgical procedures in the world. Although there is evidence that the first circumcision was performed in Egypt in 4000 BC, according to some anthropologists, it dates back to the 10th millennium BC. The purpose of medical circumcision is to obtain enough foreskin to expose the glans penis and to prevent medical problems caused by the foreskin. Although it is known that the complications arising from these procedures are not well documented, the complication rates in the literature vary between 1 and 15%, when evaluated according to age, the rate of post-circumcision complications in newborns is reported to be approximately 0.2-0.6% and this rate is 10 times higher between the ages of 1-9. Various complications such as bleeding, infection, incomplete and insufficient circumcision, hematoma, penile adhesion, urinary retention, glanular injury, necrosis and urethral narrowing have been reported. In this book section, one of the complications, penile necrosis, will be explained in the light of the literature.

Keywords: Circumcision, Glans Necrosis, Penile Necrosis, Complication, surgical technique

1. Introduction

Male circumcision is among the most commonly performed procedure for traditional, religious, and medical reasons in the world [1]. Circumcision is the process of surgically cutting the foreskin, called the prepuce, covering the glans penis, in a certain shape and length, and exposing the tip of the penis. While it is estimated that 25% of men are circumcised all over the world, this rate is 99% in Turkey and Africa [2, 3]. Circumcision operations, which have been performed in many societies from the past to the present, are now performed by doctors from various specialties [4]. Although circumcision is not a simple procedure, it is still widely performed by insufficient educated, uneducated, and unlicensed individuals. Recently, with the increasing awareness of the society, the number of circumcisions performed by traditional circumcisers tended to decrease and the number of applications made to specialist surgeons increased significantly [3, 5].

Many techniques have been described for the circumcision operation. The main purpose of circumcision is to achieve a better cosmetic result [6]. In the literature, complication rate is between 1 and 15%nafter circumcision. According to age assessments, the incidence of complications after circumcision in newborns according to age assessment is reported as approximately 0.2-0.6% and this ratio is more than 10 times in the range of 1 to 9 years [7, 8]. If surgical procedures for circumcision are performed step by step, complications are extremely rare. However, complications such as bleeding, infection and scar formation can still be seen. Apart from these complications, complications such as penile necrosis, which are much rarer but difficult to treat and manage, can also be seen [3, 5].

The most common causes of penile necrosis after circumcision include dorsal nerve block with local anesthetic agents, increased compression of the wound dressing, and the use of inappropriate surgical techniques and devices [9]. Although there are various medical and surgical treatment options in the literature in patients with penile necrosis, there is still no consensus on treatment management. Data in the literature on penile necrosis could not go further than case series. The aim of this study is to provide a medicolegal presentation of the findings and results acquired in the evaluation of a high volume of cases for the first time.

2. Circumcision time

It has been widely reported that as the age of the patient increases, complications of circumcision occur more frequently. Bleeding is more common in the "mini puberty" period of infancy that starts at 4 weeks and extends to 3 months. This is thought to be due to hormonally-mediated increase in penis and foreskin size and vascularization [10]. Banieghbal reported only two minor bleeding complications requiring suturing in a recent prospective, observational study of 583 neonatal circumcisions. Both occurred in 3-week-old infants. He reported that, based on the use of the Neonatal Pain Scale, the ideal framework for 'painless' circumcision is first week of life [11]. Horowitz and Gershbein found no complications in 98 babies circumcised with the Gomco clamp in the first month of life and supported neonatal circumcision. However, when they circumcised 3-8 months old children with the Gomco clamp, they found the rate of bleeding requiring suture or fulguration at a rate of 30% [12].

3. Circumcision techniques

3.1 Dorsal-ventral slit technique

After preparing the penis and foreskin, the devascularized tissue is cut with the help of a clamp on the dorsal part. Although the glans becomes visible after this application, it is generally not preferred alone because a good cosmetic result is not obtained and continues with the foreskin excision. Similar situation made from ventral face also applies to the operation [13].

3.2 Sleeve technique

The skin of the penis is incised by marking at a sufficient length. A second incision is made by retracting the foreskin and marking the mucosa just below the coronal sulcus. The tissue in between is removed. Bleeding control is achieved with the help of cautery or by tying. Wound lips are closed with absorbable sutures [14].

3.3 Guillotine

The foreskin is suspended from the dorsal and ventral side with the help of a clamp. The areas previously marked on the foreskin are pulled towards the glans tip. The foreskin is clamped with a straight clamp. While doing this, make sure that the glans is not compressed. The foreskin remaining distal to the clamp is cut with a scalpel and the clamp is removed. After bleeding control, the ends are tied with absorbable sutures [15].

3.4 Special circumcision clamps (Gomco clamp, Mogen clamp, Plastibell, etc.)

It is mostly preferred in newborn circumcisions. In the United States, the Gomco clamp, plastibel and Mogen clamp are mostly used. The procedure is similar for the Gomco clamp and the plastic sleeve. After the necessary penile preparation is made, a dorsal slit is made. The bell part of the Gomco clamp is placed over the glans. The other plane of the clamp is removed from the hole in the foreskin to the marked place and the screw is tightened. After waiting for a few minutes, the foreskin is cut out from the distal of the clamp with the help of a scalpel (cautery should not be used). The clamp is opened and removed from the penis. Usually the edges are squeezed together so there is no need for bleeding control and suturing. The application in the Mogen clamp is generally similar. However, dorsal slit is not required in this method [16, 17].

3.5 Circumcision with diathermic knife

This method, which has been brought to the literature by us especially with its application in bleeding diathesis, consists of the combination of the guillotine principle and a special clamp and a cutting tool that delivers thermal energy.

First, the penis is cleaned and the foreskin is separated from the glans. Similar to the guillotine method, the foreskin is suspended from the dorsal and ventral side with the help of a clamp. The foreskin is clamped with the help of a clamp prepared for this purpose, so that it does not get into the glans. The foreskin is cut out from the distal of the clamp with the help of a thermo-cautery (diathermic knife). The clamp is opened and the ends are sutured with absorbable sutures. This method does not require additional bleeding control. There is no electric current at the tip of the thermo-cautery used, and it works with thermal principles [18].

As a special case, children with bleeding disorders should not be circumcised if possible. However, this is often not possible in societies such as our country, where circumcision is seen as a requirement of masculinity. While these children should be circumcised by a physician under appropriate conditions, in some cases, families may risk having the child circumcised without taking adequate precautions and have serious bleeding problems. The current treatment principles of bleeding disorders include the social and cultural integration of patients into the society. For this purpose, it is recommended to provide circumcision under safe conditions for patients who request [18].

4. Circumcision complications

Indications for circumcision generally develop due to medical, religious and cultural reasons. Post-circumcision complications are classified as early and late complications. Early complications are bleeding, pain, inadequate skin excision, surgical site infection, and these complications are usually treated conservatively or medically. Chordee, iatrogenic hypospadias, penile necrosis and glandular amputation are among the serious early complications. Late complications include epidermal inclusion cysts, urinary retention, skin bridge formation between the penile shaft and glans, cordial, penile adhesions, phimosis, hidden penis, urethro-cutaneous fistula, and meatal strictures. Most of these complications can be avoided by paying attention to surgical principles and specific techniques used [19, 20].

5. Glans and penile necrosis

Among these complications, glans ischemia or necrosis is rarely seen. Although not all causes of post-circumcision necrosis are presented here, this is multifactorial In the pathogenesis of necrosis, the use of vasoconstrictive agents containing local anesthetic drugs, certain surgical methods, attachment of vascular structures during the procedure, excessive monopolar cautery use, wound bandage compression during nocturnal penile erections and post-circumcision infections are blamed [21, 22]. Circumcision is a surgical procedure and therefore should only be done by a doctor. Today, it is known that it is done by unlicensed circumcisers in most countries [7]. Complications are more common in circumcisions performed by unlicensed circumcisers. While the complication rate is 1% in circumcisions performed by specialist physicians, it reaches 10% when performed by unlicensed health technicians and reaches 85% when performed by traditional circumcisions [3].

5.1 Etiological factors

5.1.1 Local anesthesia

Peripheral nerve blocks are one of the most commonly used regional anesthesia methods. It allows many surgical procedures to be performed painlessly. The analgesia obtained by injecting the local anesthetic solution into the target tissue peripheral nerve or nerve plexus bed occurs simply by blocking sodium transport across the neuron cell membrane. Today, the increasing demand for circumcision under local anesthesia has increased the frequency of local anesthesia use in children [23].

Although rarely encountered, accidental injection of high doses of local anesthetic into the systemic circulation or tissue spaces can lead to serious complications. The most important of these, and the most common cause of mortality, local anesthetic toxicity is 20 to 25 minutes of the drug injected into the tissue bed at a dose higher than the safe limit. It occurs with the systemic circulation in it [2]. Its main target is the central nervous system (CNS) and cardiovascular system (CVS), whose blood supply is primary in the body. The more sensitive CNS is affected early and first manifests with restlessness (agitation, confusion) and tinnitus. Later, non-specific findings such as metallic taste in the mouth, perioral numbness, speech impairment, diplopia and convulsions are observed [24].

Local anesthesia complications that may occur in operating rooms can be detected by closely monitoring. These complications can be eliminated in the early period with appropriate medication and anesthetic support. However, the same situation may not be the case where there is no suitable equipment and support, and the unwanted consequences that may occur may bring heavy legal and conscientious burdens [25].

One of the factors that trigger the development of necrosis is the content of local anesthetic agents. Especially local anesthetic drugs containing vasoconstrictor agent reduce our blood flow to the tissue and may cause ischemia [26]. The circumcision operation is performed under local anesthesia by dorsal nerve blockage. Anesthetic drugs such as ropivacaine and lidocaine are most commonly used for dorsal nerve blockade, and when these agents are combined with vasoconstrictor drugs such as ephedrine and epinephrine, the probability of developing glans necrosis increases [27]. Apart from the vasoconstrictor effect of local anesthesia, endothelial damage due to needle insertion during dorsal nerve blockage may cause vasospasm and this may cause glans necrosis [21]. Cytochrome b5 reductase activity is lower in newborns and premature infants compared to adults. Considering that drugs such as lidocaine are metabolized by this enzyme, it should be kept in mind that methemoglobinemia may develop in children after dorsal nerve blockage [28]. The drugs most commonly used in dorsal nerve block are Prilocaine and lidocaine, and the maximum dose is 1-2 mg/kg for prilocaine and 3 mg/kg for lidocaine. These doses may produce low levels of methemoglobin, which does not cause cyanosis, which will not cause cyanosis, but the risk of methemoglobinemia increases parallel to dose increases [29].

5.1.2 Surgical technique

Circumcision surgery can be performed with many methods and modifications of these methods. Open surgical methods such as guillotine technique, dorsal slit or Sleeve technique and special clamping methods such as Mogen clamp, Gomco clamp, Plastibell device are basic surgical procedures. In addition, less common techniques such as thermal energy or laser cutting or using adhesives instead of stitches have been also reported [30, 31]. Regarding the methods used, glans necrosis generally occurs as a result of damage caused by cautery used during circumcision performed with a Gomco clamp or a wrong-sized Plastibell ring placed on the glans penis [32, 33].

5.1.3 Cautery type

It is known that the cautery type used for bleeding control during circumcision also plays a significant role in complication development. Thermal methods are frequently used in energy sources used in surgical hemostasis and one of them is electrocautery. Electrocautery instruments are of 2 types, monopolar and bipolar. In monopolar cauterization, the current is between 2 distant electrodes and the patient himself is a complement to the circuit. Energy exits the active small electrode and leaves the body through the large collecting electrode attached to the patient's skin. There are some risks and dangers associated with this circuit system in the use of monopolar energy. The most common complication is burns [34]. In the bipolar system, without the use of a cautery plate, the only instrument contains both active and collecting electrodes. High frequency electric current passes between two electrodes whose ends are close to each other. The tissue between the two electrodes is the tissue to which electrosurgery is targeted. In bipolar, the effect on the tissue is created by using less energy than monopolar. This results in fewer potential injuries [35, 36].

Bipolar energy devices can be used safely for hemostasis in penile surgeries. In monopolar energy sources, the electric current can be transferred to the small penis hole and spread to the entire penis; thermal damage may occur as a result [37]. A case of postoperative penile necrosis due to monopolar cautery used during circumcision has been reported from Turkey [38].

5.1.4 Infection

The largest study on the complications of circumcision in the literature was conducted with 100.157 pediatric patients, and the rate of infection after circumcision was reported as 0.06% in this study. Poorly sterilized surgical instruments, insufficient sterilization of the surgical field, and poor wound dressing have been shown to cause this situation. Considering these penile necrosis risk factors, the possibility of these complications, which are severe and difficult to reverse, should be decreased both before, during, and after surgery [39].

5.2 Diagnosis

Penile necrosis is diagnosed with physical examination findings. Radiology can be assessed visually using Doppler USG for the degree of ischemia and necrosis [40]. Modern sonography equipment with high-frequency probes up to 15 MHz provides detailed visualization of penile anatomy and vascular structure in normal and pathological conditions, which are of great value in the evaluation of penile hemodynamics in patients. It should be noted that the glans skin is less vascularized than the highly vascular corpora cavernosa, and blood flow velocities and volumes in the glant may be below the threshold of a scanner [41].

Doppler USG is used not only during diagnosis but also during follow-up. Doppler USG is helpful in determining to what extent the degree of necrosis is restored with treatment. Penile necrosis may require rapid penile amputation after diagnosis or cause post-operative uromechanical complications (such as urethral stenosis and meatal stenosis) and late sexual dysfunction. Prompt treatment is essential after diagnosis to minimize these complications [40, 41].

5.3 Treatment

The main goal in the treatment of glans necrosis is to increase tissue vascularization and oxygenate ischemic penile tissue [42]. In the literature, different methods have been tried in the treatment of glans necrosis, but there is no consensus on the treatment. The pharmacological agents commonly used today are pentoxifylline (PTX), enoxaparin, iloprost, antiplatelet drugs, corticosteroids and caudal anesthesia [21, 22].

In previous studies, it was stated that the use of oral PTX will provide successful results when used in the treatment of post-circumcision glans ischemia due to the effect of prostaglandin production on ischemia–reperfusion injury and the inhibitory effect of phosphodiesterase activity causing cyclic adenosine monophosphate synthesis [22]. Other studies in the literature have reported successful use of ropivacaine, subcutaneous enoxaparin, and topical testosterone in the treatment of penile necrosis [17, 22, 43].

Another treatment method used as an alternative to medical treatment is HBOT, which is suggested as an adjuvant treatment method because it increases tissue oxygenation, neutralizes anaerobic bacteria, improves neutrophil functions, increases fibroblast proliferation, and stimulates angiogenesis [44].

When the literature is reviewed, various pharmacological agents have been used in the treatment of glans and penile necrosis after circumcision. Aminsharifi et al. [45] treated their patients using topical 10% testosterone undecanoate twice a month and found significant improvement in their patients. Ozzeybek et al. [46] used glycerol trinitrate 3 mg/kg, 2 mL) as intracavernous and Bupivacaine 0.0625% as epidural for 5 days and they saw complete improvement after 1 week. Burke et al. [27] administered intravenous iloprost (PGI2 analogue) at 0.52 μg/h to a patient who developed glans necrosis after dorsal penile block and they observed complete recovery after 43 hours. Sara and Lowry administered low-dose molecular weight heparin infusion (25 units/kg/h) for 4 days. Tzeng *et al.* and Aslan *et al.*, [21, 22] and more recently Elemen *et al.* [47] reported 2 patients of 5 and 33 years old treated with intravenous pentoxifylline (10 mg/kg) divided in four daily doses, together with hyperbaric oxygen with 2.5 atm of pressure for 90 min-long sessions. Total reverse of the ischemia was observed in both cases.

In addition to these treatments, necrotic tissue debridement and graft application constitute one of the main treatment methods [48]. Through the debridement of necrotic tissue, the progress of the disease is quickly stopped, and successful results are achieved. In some studies, it was reported that severe cases could be successfully managed with suprapubic diversion and delayed urethroplasty [32]. Rare cases have been reported of complete glans and phallus necrosis following multiple interventions in staged repairs [33].

6. Medicolegal aspect

As in all surgical procedures, it is the realistic and comprehensive preoperative information provided by the relevant surgeon to increase the patient (parent) satisfaction after the circumcision operation. Informed consent, if available in a standard form, should be obtained with a signature explaining the surgical technique and listing all possible complications of the operation. The consent form should include the consent of the patient/parent for circumcision, the indications for circumcision in the individual case, surgical technique, and complications that may develop in the postoperative period. Christensen-Szalamski et al. [49] showed that replacing the `physician's policy of partial disclosure with a comprehensive disclosure of unbiased information of possible risks and complications' had no effect on the mother's decision to have their son circumcised. If the procedure is performed under local anesthesia applied by the surgeon, the risks of this should also be discussed. If a child is to be circumcised, it is desirable to get his consent by explaining the procedure in accordance with his cognitive abilities.

From a medical law perspective, a retrospective review of malpractice in urology in a study conducted in the U.S. found that only 5.2% of cases were circumcision-related (adult or pediatric). Similarly, the largest jury review of urology malpractice cases showed that 3% of cases were circumcision related [50]. In the UK, the circumcision operation can only be legally performed by doctors [51]. In Turkey and Middle East, the Ministry of Health certifies those who attend a three-month course and who prove that they have been circumcised for 10 years even if there are no health personel. Those who are convicted of performing unlicensed circumcision generally receive a light prison sentence of one week to one month [52].

7. Conclusion

According to the literature data, the lifelong medical benefits of circumcision of boys outweigh the risks of circumcision. Although there are contradictory results about the effect of circumcision on mental health, it is seen that it varies according to culture and affects mental health positively in societies where circumcision is common. Therefore, circumcision is beneficial in terms of biological and mental health, if performed under recommended age and conditions.

As an operation commonly performed in many countries and increasingly preferred due to medical and social reasons, circumcision brings different complications. More importantly, most of these complications can be prevented by careful surgery and careful postoperative care. Although the complications are generally minor, this surgical procedure is still performed by individuals other than doctors, and the tradition of mass circumcision causes major complications. It suggests that the rate of complications will decrease if the physicians who have received circumcision penile surgery education do it.

Glans necrosis is a rare complication but causes serious cosmetic, reproductive, and psychological problems after circumcision and may occur after cautery. The preferred treatment for glanular necrosis has not been established yet, different treatment protocols are specified in the literature. The most important point in the treatment of glans necrosis is early diagnosis and rapid intervention.

Conflict of interest

All listed authors declare that he/she has no conflict of interest.

Author contributions

Study conception and design (YA, AA), Data acquisition (YA, AA), Analysis and data interpretation (YA, AA), Drafting of the manuscript (YA, AA), and Critical revision (YA, AA).

Author details

Yusuf Arikan* and Ali Ayten
Department of Urology, Bakırkoy Dr. Sadi Konuk Training and Research Hospital, Istanbul, Turkey

*Address all correspondence to: dryusufarikan@gmail.com

IntechOpen

© 2021 The Author(s). Licensee IntechOpen. This chapter is distributed under the terms of the Creative Commons Attribution License (http://creativecommons.org/licenses/by/3.0), which permits unrestricted use, distribution, and reproduction in any medium, provided the original work is properly cited.

References

[1] Dunsmuir WD, Gordon EM. The history of circumcision. BJU Int. 1999 Jan;83 Suppl 1:1-12. doi: 10.1046/j.1464-410x.1999.0830s1001.x. PMID: 10349408.

[2] Morris BJ, Wamai RG, Henebeng EB, Tobian AA, Klausner JD, Banerjee J et al. Estimation of country-specific and global prevalence of male circumcision. Popul Health Metr. 2016 Mar 1;14:4. doi: 10.1186/s12963-016-0073-5. eCollection 2016.

[3] Ozdemir E. Significantly increased complication risks with mass circumcisions. Br J Urol. 1998 Apr;81(4):652. PMID: 9598657.

[4] Moses S, Bailey RC, Ronald AR. Male circumcision: assessment of health benefits and risks. Sex Transm Infect. 1998 Oct;74(5):368-73. doi: 10.1136/sti.74.5.368. PMID: 10195035; PMCID: PMC1758146.

[5] Morris BJ, Moreton S, Krieger JN. Critical evaluation of arguments opposing male circumcision: A systematic review. J Evid Based Med. 2019 Nov;12(4):263-290. doi: 10.1111/jebm.12361. Epub 2019 Sep 8. PMID: 31496128; PMCID: PMC6899915.

[6] Kaplan GW. Complications of circumcision. Urol Clin North Am. 1983 Aug;10(3):543-549. PMID: 6623741.

[7] Harrison NW, Eshleman JL, Ngugi PM. Ethical issues in the developing world. Br J Urol. 1995 Nov;76 Suppl 2:93-96. doi: 10.1111/j.1464-410x.1995.tb07879.x. PMID: 8535764.

[8] Weiss HA, Larke N, Halperin D, Schenker I. Complications of circumcision in male neonates, infants and children: a systematic review. BMC Urol. 2010 Feb 16;10:2. doi: 10.1186/1471-2490-10-2. PMID: 20158883; PMCID: PMC2835667.

[9] Hornez E, Laroche J, Monchal T, Bourgouin S, Riviere P, Fournier R, Dantzer E. Necrosis of the glans penis: a complication of an injection of buprenorphin in a opioid abuser. Ann Chir Plast Esthet. 2010 Apr;55(2):159-61. French. doi: 10.1016/j.anplas.2008.12.004. Epub 2009 Mar 9. PMID: 19269730.

[10] Blank S, Brady M, Buerk E, Carlo W, Diekema D, Freedman A et al. American Academy of Pediatrics Task Force on Circumcision. Male circumcision. Pediatrics. 2012 Sep;130(3):e756-e785. doi: 10.1542/peds.2012-1990. Epub 2012 Aug 27.

[11] Banieghbal B. Optimal time for neonatal circumcision: an observation-based study. J Pediatr Urol. 2009 Oct;5(5):359-362. doi: 10.1016/j.jpurol.2009.01.002. Epub 2009 Feb 14. PMID: 19223238.

[12] Horowitz M, Gershbein AB. Gomco circumcision: When is it safe? J Pediatr Surg. 2001 Jul;36(7):1047-1049. doi: 10.1053/jpsu.2001.24739. PMID: 11431774.

[13] Lei JH, Liu LR, Wei Q, Xue WB, Song TR, Yan SB, Yang L, Han P, Zhu YC. Circumcision with "no-flip Shang Ring" and "Dorsal Slit" methods for adult males: a single-centered, prospective, clinical study. Asian J Androl. 2016 Sep-Oct;18(5):798-802. doi: 10.4103/1008-682X.157544. PMID: 26585694; PMCID: PMC5000807.

[14] Jiang ZL, Sun CW, Sun J, Shi GF, Li H. Subcutaneous tissue-sparing dorsal slit with new marking technique: A novel circumcision method. Medicine (Baltimore). 2019 Apr;98(16):e15322. doi: 10.1097/MD.0000000000015322. PMID: 31008987; PMCID: PMC6494260.

[15] Basar H, Yilmaz E, Basar MM, Batislam E, Tuglu D. Window technique

on circumcision. Int Urol Nephrol. 2006;38(3-4):599-601. doi: 10.1007/s11255-005-0249-6. PMID: 17033889.

[16] Nicassio L, Ching CB, Klamer B, Sebastião YV, Fuchs M, McLeod DJ, Alpert S, Jayanthi R, DaJusta D. Gomco vs. plastibell office circumcision: No difference in overall post-procedural complications and healthcare utilization. J Pediatr Urol. 2021 Feb;17(1):85.e1-85.e7. doi: 10.1016/j.jpurol.2020.11.019. Epub 2020 Nov 15. PMID: 33281046.

[17] Chan PS, Penna FJ, Holmes AV. Gomco Versus Mogen? No Effect on Circumcision Revision Rates. *Hospital Pediatrics.* 2018;8(10):611-614. Hosp Pediatr. 2019 Mar;9(3):225. doi: 10.1542/hpeds.2018-0263. Erratum for: Hosp Pediatr. 2018 Oct;8(10):611-614. PMID: 30819721.

[18] Karaman MI, Zulfikar B, Özturk MI, Koca O, Akyüz M, Bezgal F. Circumcision in bleeding disorders: improvement of our cost effective method with diathermic knife. Urol J. 2014 May 6;11(2):1406-1410. PMID: 24807751.

[19] Krill AJ, Palmer LS, Palmer JS. Complications of circumcision. ScientificWorldJournal. 2011;11:2458-68. doi: 10.1100/2011/373829. Epub 2011 Dec 26. PMID: 22235177; PMCID: PMC3253617.

[20] Douglas M, Maluleke TX, Manyaapelo T, Pinkney-Atkinson V. Opinions and Perceptions Regarding Traditional Male Circumcision With Related Deaths and Complications. Am J Mens Health. 2018;12(2):453-462. doi:10.1177/1557988317736991

[21] Tzeng YS, Tang SH, Meng E, Lin TF, Sun GH. Ischemic glans penis after circumcision. Asian J Androl. 2004 Jun;6(2):161-163. PMID: 15154092.

[22] Aslan A, Karagüzel G, Melikoglu M. Severe ischemia of the glans penis following circumcision: a successful treatment via pentoxifylline. Int J Urol. 2005 Jul;12(7):705-707. doi: 10.1111/j.1442-2042.2005.01129.x. PMID: 16045570.

[23] Dickerson DM, Apfelbaum JL. Local anesthetic systemic toxicity. Aesthet Surg J. 2014 Sep;34(7):1111-1119. doi: 10.1177/1090820X14543102. Epub 2014 Jul 15. PMID: 25028740.

[24] Felice K, Schumann H. Intravenous lipid emulsion for local anesthetic toxicity: a review of the literature. J Med Toxicol. 2008 Sep;4(3):184-91. doi: 10.1007/BF03161199. PMID: 18821493; PMCID: PMC3550038.

[25] Barrington MJ, Kluger R. Ultrasound guidance reduces the risk of local anesthetic systemic toxicity following peripheral nerve blockade. Reg Anesth Pain Med. 2013 Jul-Aug;38(4):289-299. doi: 10.1097/AAP.0b013e318292669b. PMID: 23788067.

[26] Gupta R, Garg M, Pawah S, Gupta A. Postanesthetic ulceration of palate: A rare complication. Natl J Maxillofac Surg. 2016 Jan-Jun;7(1):86-88. doi: 10.4103/0975-5950.196142. PMID: 28163486; PMCID: PMC5242082.

[27] Burke D, Joypaul V, Thomson MF. Circumcision supplemented by dorsal penile nerve block with 0.75% ropivacaine: a complication. Reg Anesth Pain Med. 2000 Jul-Aug;25(4):424-427. doi: 10.1053/rapm.2000.7594. PMID: 10925943.

[28] Guay J. Methemoglobinemia related to local anesthetics: a summary of 242 episodes. Anesth Analg. 2009 Mar;108(3):837-845. doi: 10.1213/ane.0b013e318187c4b1. PMID: 19224791.

[29] Rosen M. Anesthesia for ritual circumcision in neonates. Paediatr Anaesth. 2010 Dec;20(12):1124-1127.

doi: 10.1111/j.1460-9592.2010.03445.x. PMID: 21199122.

[30] American Academy of Pediatrics Task Force on Circumcision. Male circumcision. Pediatrics. 2012 Sep;130(3):e756-e785. doi: 10.1542/peds.2012-1990. Epub 2012 Aug 27. PMID: 22926175.

[31] Elemen L, Seyidov TH, Tugay M. The advantages of cyanoacrylate wound closure in circumcision. Pediatr Surg Int. 2011 Aug;27(8):879-883. doi: 10.1007/s00383-010-2741-z. Epub 2010 Oct 13. PMID: 20941598.

[32] Bode CO, Ikhisemojie S, Ademuyiwa AO. Penile injuries from proximal migration of the Plastibell circumcision ring. J Pediatr Urol. 2010 Feb;6(1):23-27. doi: 10.1016/j.jpurol.2009.05.011. Epub 2009 Jun 30. PMID: 19570722.

[33] Gee WF, Ansell JS. Neonatal circumcision: a ten-year overview: with comparison of the Gomco clamp and the Plastibell device. Pediatrics. 1976 Dec;58(6):824-827. PMID: 995507.

[34] Advincula AP, Wang K. The evolutionary state of electrosurgery: where are we now? Curr Opin Obstet Gynecol. 2008 Aug;20(4):353-358. doi: 10.1097/GCO.0b013e3283073ab7. PMID: 18660686.

[35] Rioux JE. Bipolar electrosurgery: a short history. J Minim Invasive Gynecol. 2007 Sep-Oct;14(5):538-541. doi: 10.1016/j.jmig.2007.06.007. PMID: 17848309.

[36] Barrett SL, Vella JM, Dellon AL. Historical development of bipolar coagulation. Microsurgery. 2010 Nov;30(8):667-669. doi: 10.1002/micr.20815. Epub 2010 Sep 14. PMID: 20842704.

[37] Harty NJ, Nelson CP, Cendron M, Turner S, Borer JG. The impact of electrocautery method on post-operative bleeding complications after non-newborn circumcision and revision circumcision. J Pediatr Urol. 2013 Oct;9(5):634-637. doi: 10.1016/j.jpurol.2012.06.019. Epub 2012 Aug 1. PMID: 22858383.

[38] Uzun G, Ozdemir Y, Eroglu M, Mutluoglu M. Electrocautery-induced gangrene of the glans penis in a child following circumcision. BMJ Case Rep. 2012 Oct 29;2012:bcr-2012-007096. doi: 10.1136/bcr-2012-007096. PMID: 23109415; PMCID: PMC4544066.

[39] Wiswell TE, Geschke DW. Risks from circumcision during the first month of life compared with those for uncircumcised boys. Pediatrics. 1989 Jun;83(6):1011-1015. PMID: 2562792.

[40] Barnes S, Ben Chaim J, Kessler A. Postcircumcision necrosis of the glans penis: gray-scale and color Doppler sonographic findings. J Clin Ultrasound. 2007 Feb;35(2):105-107. doi: 10.1002/jcu.20271. PMID: 17195193.

[41] du Toit DF, Villet WT. Gangrene of the penis after circumcision: a report of 3 cases. S Afr Med J. 1979 Mar 24;55(13):521-522. PMID: 451776.

[42] Sterenberg N, Golan J, Ben-Hur N. Necrosis of the glans penis following neonatal circumcision. Plast Reconstr Surg. 1981 Aug;68(2):237-239. doi: 10.1097/00006534-198108000-00022. PMID: 7255584.

[43] Efe E, Resim S, Bulut BB, Eren M, Garipardic M, Ozkan F, Ozkan KU. Successful treatment with enoxaparin of glans ischemia due to local anesthesia after circumcision. Pediatrics. 2013 Feb;131(2):e608-e611. doi: 10.1542/peds.2012-1400. Epub 2013 Jan 14. PMID: 23319528.

[44] Willy C, Rieger H, Vogt D. Hyperbare Oxygenation bei nekrotisierenden Weichteilinfektionen:

Kontra [Hyperbaric oxygen therapy for necrotizing soft tissue infections: contra]. Chirurg. 2012 Nov;83(11):960-72. German. doi: 10.1007/s00104-012-2284-z. PMID: 23138865.

[45] Aminsharifi A, Afsar F, Tourchi A. Delayed glans necrosis after circumcision: role of testosterone in salvaging glans. Indian J Pediatr. 2013 Sep;80(9):791-793. doi: 10.1007/s12098-012-0820-y. Epub 2012 Jun 30. PMID: 22752705.

[46] Ozzeybek D, Koca U, Elar Z, Olguner M, Hakgüder G. Glycerol trinitrate plus epidural sympathetic block in the ischemia of glans penis. Anesth Analg. 1999 Oct;89(4):1066. doi: 10.1097/00000539-199910000-00053. PMID: 10512297.

[47] Elemen L, Topçu K, Gürcan Nİ, Akay A. Successful treatment of post circumcision glanular ischemia-necrosis with hyperbaric oxygen and intravenous pentoxifylline. Actas Urol Esp. 2012;36: 200-201. doi: 10.1016/j.acuro.2011.06.008. Epub 2011 Aug 6. PMID: 21821318.

[48] Sugihara T, Yasunaga H, Horiguchi H, Fujimura T, Ohe K, Matsuda S, Fushimi K, Homma Y. Impact of surgical intervention timing on the case fatality rate for Fournier's gangrene: an analysis of 379 cases. BJU Int. 2012 Dec;110(11 Pt C):E1096-100. doi: 10.1111/j.1464-410X.2012.11291.x. Epub 2012 Jun 21. PMID: 22726768.

[49] Christensen-Szalamski JJ, Boyce WT, Harrell H, Gardner MM. Circumcision and informed consent. Is more always better? Med Care 1987; 25: 856-867.

[50] Hsieh MH, Tan AG, Meng MV. Medical malpractice in American urology: 22-year national review of the impact of caps and implications for contemporary practice. J Urol. 2008 May;179(5):1944-1949; discussion 1949. doi: 10.1016/j.juro.2008.01.061. Epub 2008 Mar 20. PMID: 18355843.

[51] Kahan SE, Goldman HB, Marengo S, Resnick MI. Urological medical malpractice. J Urol. 2001 May;165(5):1638-1642. PMID: 11342944.

[52] Beecham L. GMC issues guidelines on circumcision. BMJ. 1997 May 31;314(7094):1573. doi: 10.1136/bmj.314.7094.1569h. PMID: 9186166; PMCID: PMC2126813.

Chapter 9

Epigenetics in Male Infertility

Hayfa H. Hassani, Rakad M. Kh AL-Jumaily and Fadhel M. Lafta

Abstract

Male infertility is a complex medical condition, in which epigenetic factors play an important role. Epigenetics has recently gained significant scientific attention since it has added a new dimension to genomic and proteomic research. As a mechanism for maintaining genomic integrity and controlling gene expression, epigenetic modifications hold a great promise in capturing the subtle, yet very important, regulatory elements that might drive normal and abnormal sperm functions. The sperm's epigenome is known to be marked by constant changing over spermatogenesis, which is highly susceptible to be influenced by a wide spectrum of environmental stimuli. Recently, epigenetic aberrations have been recognized as one of the causes of idiopathic male infertility. Recent advances in technology have enabled humans to study epigenetics role in male infertility.

Keywords: Male infertility, Epigenetics, Environmental epigenetics

1. Introduction

Male infertility is a complex condition of which genetic, epigenetic and environmental lifestyle have been identified as major contributing factors [1–3]. In spermatogenesis, which is a complex of multistep differentiation process, millions of mature spermatozoa are daily produced by fertile male. In addition, this process comprises a variety of unique genetic and epigenetic mechanisms that eventually generate haploid sperm, which provides half of the genetic material and epigenetic information that is needed to create a new life upon fertilization. The sperm's differentiation is error prone and problems at all stages of spermatogenesis might contribute to male infertility [4]. It is believed that epigenetic modifications are essential to regulate normal gonadal development and spermatogenesis. This includes the normal distribution of variant epimarks controlling the testis-tissue specific chromatin compaction and the resultant gene expression accordingly. In this regards, several lines of evidence have highlighted the present of abnormal epigenetic marks in somatic and germ testicular tissues that are associated with impaired fertility or poor semen criterions. Hypermethylated of genes involve in PIWI -associated small RNAs (piRNAs) have been observed in testicular tissues of males having various forms of fertility problems [5]. Within this context, studies have reported disrupted epigenetics patterns of cells from different testicular tissues including Leydig [6] and Sertoli cells [7]. Interestingly, in patients with low testicular volume have been reported to have lower chromatin compactness and poor sperm quality [8].

IntechOpen

At present, efforts have been concentrated on understanding the potential key role for epigenetic modifications in male reproduction health and the prevention of paternal disease transmission [9]. Epigenetics is defined as molecular factors or processes around DNA that regulate germline activity independent of DNA sequence and are mitotically stable [10]. Epigenetic changes are also a set of factors that affect the expression of genes, but do not affect the DNA sequence. In this chapter, we review the epigenetic marks in normal and abnormal human sperms, the influence of environmental stimuli on germ cells' epigenetic modifications in relation to male infertility, and technologies used for the analysis of epigenetic modifications associated with male infertility.

2. Epigenetic profile in normal sperms

The chromosomes of sperms are arranged in a hairpin-like structure, with the centromeres being confined to the interior of the nucleus and the telomeres being at the periphery [11]. The DNA of sperm is packed with specific, small, basic proteins into a tight, almost crystalline, status that is at least six times more condensed than that in mitotic chromosomes [12]. In sperms, the somatic cell histones are replaced with 90–95% of specific basic nuclear proteins known as protamines, leading to highly packaged chromatin. There are two types of protamines (P1 and P2). The P1 protamine is present in all of the mammalian species while P2 promatine is a family proteins formed by the P2, P3 and P4 components present in some of the mammalian species. Actually, protamines have many functions, such as enabling faster sperm movement and, thus, having the potential to fertilize the oocyte first. In addition, they are involved in the imprinting of the paternal genome during spermatogenesis [13]. In some of the mammalian species, including humans, this replacement involves a set of special proteins, i.e. a group of arginine (R)- and lysine (K)-rich proteins, known as transition nuclear proteins (TPs) [14]. The major TPs are TP1 and TP2. TP1 play an important role in the initiation of chromatin condensation and/or cessation of transcriptional activity during mammalian spermatogenesis [12], while TP2 is closely linked to the two protamine genes [15], suggesting that they arose by gene duplication and might have retained common functions. About 5–10% of nucleohistone component retains within sperm chromatin and provide a means for further epigenetic regulation. Expression of these TPs is presumed to regulate changes in chromatin occurring as part of the condensation process [16].

Along with the proposed role for the sperm epigenome marks in shaping the embryonic development, they also could be used for male's gamete stratification based on the comparison to their normal counterparts. Such marks include histone retention and modification, and protamine incorporation into the chromatin. In addition, DNA methylation and spermatozoal non-coding RNAs appear to play important roles in the epigenetic state of mature sperms. These epigenetic marks may also reveal a historical record of spermatogenesis, future functions in embryogenesis, and fertilization [17].

2.1 Histone modifications

Histone modifications are type of epigenetic marks that can potentially be transmitted from parent to offspring [10]. During mitosis and meiosis, male germ cell DNA is packaged in nucleosomes comprised of histone 2A (H2A), histone 2B (H2B), histone 3 (H3) and histone 4 (H4), all of which are susceptible to covalent modifications, such as methylation, acetylation, ubiquitination and phosphorylation. Each of these histone chemical modifications works alone or in concert to influence gene repression and/or activation [18]. Histone methylation is controlled by histone methyltransferases (HMTs) which modifies lysine (K) residues of H3 or H4 and can promote gene activation and/or repression [19]. For example, monomethylation, dimethylation and trimethylation modifications of H3K4, H3K9 or H3K27 display tightly controlled temporal expression and ensure proper progression through spermatogenesis [20, 21]. Demethylation in histone is controlled by histone demethylases (HDMs). On the other hand, histone acetylation of lysine residues is dynamically regulated by histone acetyltransferases (HATs) and histone deacetylases (HDACs), both of which are shown to be essential for spermatogenesis [22]. Histone acetylation relaxes chromatin and promotes polymerase II (Pol II) gene transcription, whereas deacetylation is associated with gene silencing [23]. Acetylation levels on both H3 and H4 are completely removed during meiosis.

Histone phosphorylation occurs at serine (S) residues of all core histones and is generally associated with gene activation [24]. However, H2Ax phosphorylation (also known as gH2Ax) in germ cells confers the formation of Y sex body during spermatogenesis and is a marker for telomere clustering and double stranded breaks [25]. Moreover, ubiquination is another epigenetic mark in histone and its effects are dependent on the core histone modified. For example, ubiquination of H2A associates with transcriptional repression in sperm, whereas mono-ubiquination of H2B is linked to transcriptional activation in sperms [26].

Although most histones are replaced with protamines during the elongating spermatid stage, some of the modified nucleosomes escape the histone to protamine transition and, as a result, are retained in the mature sperm. The retained nucleosomes are enriched at CpG rich sequences that lack DNA methylation. The non-canonical histone H3.3 variant was shown to be abundant and trimethylated at K4 in these nucleosomes, while the canonical histones H3.1 and H3.2 are trimethylated at K27 [27]. Other non-canonical histone variants were reported previously to be present in the retained nucleosomes of the mature sperm; TH2B was observed in humans, whereas H2A-Bbd, H2AL1/L2, and H2BL1 were described in mammalian sperms [28].

2.2 DNA methylation

DNA methylation has been implicated in the development of spermatozoa and early embryos through the regulation of gene expression [29]. DNA methylation process primarily involves the addition of methyl groups to the 5′ carbon at cytosine residues preceding guanine nucleotides. These groups are linked together by phosphate bonds (CpG) utilizing a methyl donor like S-adenosylmethionine. CpG is clustered primarily in short CpG-rich DNA sequences named CpG islands. DNA methylation is catalyzed by a group of enzymes termed as DNA methyl transferases (DNMTs) that target these CpG islands. According to their structure and functions, DNMTs are divided into two major families in mammalian cells: maintenance methyltransferase (DNMT1) and *de novo* methyltransferases (DNMT3a, DNMT3b, and DNMT3L). In addition, DNA methylation can disrupt the process of transcription by inhibiting the binding of the transcriptional factors with the target sites. Also, the methylated cytosine residues act as the site for docking of various methylated DNA-binding proteins (MBD1, MBD2, MBD3, and Mecp2) that are recognized by various histone modifying enzymes like histone deacetylases (HDACs), which in turn can lead to gene repression [30]. Furthermore, methylcytosine can modify and potentially erase DNA methylation [22] by ten-eleven translocation protein 1 (TET1). TET1 belongs to a family of three proteins—namely, TET1, TET2, and TET3, that catalyze the successive oxidation of 5-methylcytosine (5mC) to 5-hydroxymethylcytosine (5hmC), 5-formylcytosine (5fC), and 5-carboxylcytosine (5caC) [31, 32].

The proper regulation of DNA *de novo* methylation and demethylation is essential for the normal function of the mature sperm and early embryo [18, 33]. Dynamic erasure and reestablishment of DNA methylation marks is catalyzed by TET dioxygenases and DNA DNMTs, respectively, which are required for the formation of mature sperm as well as for embryogenesis [34]. Methylation analysis in genome-wide studies have demonstrated that the promoters of developmental genes in sperm cells are highly hypomethylated [35]. A previous gene ontology analysis demonstrated that hypomethylation in mature sperm cells promoted developmental transcription and signaling, which is bound by the self-renewal of a network of transcription factors of human embryonic stem cells, including OCT4, SOX2, NANOG, KLF4 and FOXD3 proteins [29]. Moreover, recurrent regions of the sperm genome demonstrate high degrees of methylation, while transposons manifest lower levels of DNA methylation [36–38]. The paternally expressed human gene, MEST/PEG1, remains unmethylated throughout all stages of sperm development in the adult life [39, 40]. By contrast, in male germ cells, H19 gene is methylated prior to meiosis at the spermatogonial stage of development [41]. Moreover, reinitiation of mitotic division of male germ cells during puberty coincides with an upregulation of DNMT1 within the spermatocytes. However, the levels of DNMT1 in spermatocytes are increased during the early stage of meiosis, and reduced in pachytene stage spermatocytes [42]. The DNA methylation loss is then followed by *de novo* DNA methylation throughout all stages of sperm cell maturation, with global CpG methylation levels of 70% in fully mature human spermatozoa, yielding approximately 4% of total cytosines methylated [43]. However, DNMT3L is the only enzyme that is expressed in sperms within a low level [44]. While, DNMT3a expression is upregulated in the testis prior to birth and during early postnatal life, and DNMT3b expression is downregulated during embryonic development and upregulated postnatally [45].

2.3 Non-coding RNA

Regarding the non-coding RNAs as germ cells epigenetic marks, there are several ncRNAs that contribute to the transgenic epigenetics in the developing embryo. Among these are the small non-coding (sncRNA) and the long non-coding RNAs (lncRNAs).

Small ncRNAs are involved in the control of sperm production. Male germ lineages express classes of sncRNAs, including Dicer-dependent microRNAs (miRNAs), tiRNAs (tiny), as well as Dicer-independent PIWI-interacting RNAs (piRNAs) [46]. miRNAs observed in spermatozoa, such as miR-122, has been described to participate in the post-transcriptional down-regulation of the transition protein 2 (TNP2) during spermatogenesis [47]. TNP2, together with transition protein 1 (TNP1), transitionally substitute some of the histones during spermiogenesis. This intermediate step precedes and facilitates protamine replacement that is required to compact the paternal genome into the relatively small sperm head [48]. Piwi-interacting RNAs (26–31 nucleotides) are specifically expressed in the gonads, and are thought to silence transposable elements, especially in the germline, protecting the integrity of the genome. piRNAs are involved in epigenetic inheritance and they mediate their effect through PIWI-proteins, a subfamily of the argonaute family of proteins.

Another kind of sncRNA, known as Piwi-interacting RNAs (piRNAs), are highly expressed in germ cells and required for male fertility [49]. Mechanistically, piRNAs repress gene expression at both transcriptional level, by promoting *de novo* DNA methylation [50], and post-transcriptional level, by cleaving target transposon mRNAs [51].

Furthermore, lncRNAs is dynamically regulated during male germline development and may function to regulate gene expression at both transcriptional and posttranscriptional levels via genetic and epigenetic mechanisms. In the testis, expression profiles of lncRNAs can be detected at different ages and stages of development [52]. The Y- linked lncRNAs showed higher expression in pachytene spermatocytes. In human mature sperms, an association between lncRNA expression and sperm motility was indicated [53]. Mammals exhibit condensation and remodeling of their chromatin material during late spermatogenesis by omitting excess cytoplasm and replacing histones with protamine for spermatid individualization, leading to a highly compact sperm nucleus [54]. Enhancer-associated lncRNAs participate in transcriptional activation by acting over long distances on distal promoters, associated with protein factors, and by the modulation of chromatin structures [55].

The N6-methyladenosine (m6A) modification is the most prevalent internal RNA modification that has been defined as another important epigenetic and epitranscriptomic marker in eukaryotes [56]. m6A has notable influence on the regulation of gene expression at the post-transcriptional level, animal development, and human diseases [57]. The m6A modification mainly occurs in intragenic regions including coding sequences (CDS), stop codon flanking regions and 3′-UTR, especially the 3′-end of CDS and the first quarter of the 3′-UTR, also near transcription starting sites (TSS) [58], and the modification site is often on the conserved sequence of RRACH (R = A or G; H = A, C, or U) [59]. The m6A modification process is dynamically reversible and regulated by three kinds of regulatory proteins: methyltransferases (writers) such as methyltransferase like protein 3 (METTL3), methyltransferase like protein 14 (METTL14), Wilm's tumor-associated protein 1 (WTAP), Vir-like m6A methyltransferase-associated (VIRMA; also known as KIAA1429), RNA binding motif protein 15 (RBM15), and zinc finger CCCH domain protein 13 (ZC3H13); demethylases (erasers), such as obesity-associated protein (FTO) and AlkB family homolog 5 (ALKBH5); and m6A binding proteins (readers), such as the YTH domain family proteins (YTHDFs) and YTH domain containing protein 1–2 (YTHDC1–2), the insulin-like growth factor 2 mRNA binding proteins (IGF2BPs), heterogeneous nuclear ribonucleoprotein A2B1 (HNRNPA2B1), and eukaryotic translation initiation factor 3 (eIF3) [60].

A number of findings demonstrating the significance of m6A in male fertility and spermatogenesis. Methyltransferases (writers) especially METTL3 and METTL14 are widely expressed in different tissues in mouse and human, and essential for male fertility and spermatogenesis [61]. It found an ablation of *Mettl3* in germ cells severely inhibited spermatogonial differentiation and blocked the initiation of meiosis. Analysis of transcriptome and m6A profiling revealed that gene functioning in spermatogenesis had altered profiles of expression and alternative splicing [62]. Another study has proved that the lack of m6A by germ cell-specific inactivation of *Mettl3* or *Mettl14* results in spermatogonial stem cells (SSC) depletion due to significant changes in translational efficiency (TE) while double deletion of Mettl3 and Mettl14 in advanced germ cells leads to impaired spermiogenesis due to altered TE of m6A-containing transcripts [63].

On the other hand, ALKBH5 as demethylase is expressed in testes. It proved that m6A was increased in male mice with ALKBH5-targeted deletion, and the number of sperm released and incised caudal epididymis was significantly reduced, sperm morphology was abnormal and motility was greatly reduced. The results showed that fertility was impaired due to the abnormal apoptosis and production of a small number of abnormal spermatozoa during meiosis [64]. Concerning FTO as another demethylase, it was reported that FTO expression alters by using Di-(2-Ethylhexyl) phthalate, and results in an increasing m6A RNA modification, deteriorates testicular histology, reduces testosterone concentration, down-regulates spermatogenesis inducer expression, enhances oxidative stress, and increases testicular cell apoptosis [65]. YTHDC2 is one of m6A binding proteins (readers), that highly expressed in mammalians testes [66]. YTHDC2-knockout mice showed defects in spermatogenesis and the germ cells do not develop to the zygotic stage [67]. The regulation of m6A transcription by YTHDC2 is the key to the success of meiosis in the mammalian germline.

3. Aberrant sperm's epigenetic modifications

Epigenetic modification, especially DNA methylation, plays an important role in determining the differentiation potential of mammalian cells and ensuring the normal development. The methylation changes of a specific gene in the parent sperm cannot directly lead to the defective phenotype of the offspring, but it is undeniable that the methylation of CpG islands affects a series of cascade gene expressions, resulting in abnormal gene crossover network expression [68]. The potential roles of epigenetic processes in the progress of sperm function and male fertilization were studied by many researchers. This includes studies of DNA methylation, histone modifications, chromatin remodeling, and ncRNAs roles in the development of gonads and spermatogenesis [3, 69]. Recently, most of studies have been concentrated on understanding the role of different epigenetics mechanisms in male reproduction health and spermatogenesis. Infertility is the inability of males and females to achieve a pregnancy after 12 months or more by natural means. The development and normal functioning of male reproductive system are thought to be highly sensitive to epigenetic changes. However, the epigenetic modifications underlying male infertility remain unclear [29]. The role of epigenetic modifications in infertility and their impact on male reproduction are summarized in this review.

3.1 Aberrant sperm's DNA methylation patterns

DNA methylation plays critical roles in the regulation of gene expression during the development of mature spermatozoa, which may lead to male infertility [69, 70]. The process of germ cell development is highly organized, starting from early fetal life and is finished in the adulthood. The epigenetic alterations occurring in germ cells, including DNA methylation, are significant for the healthy development of sperm function and for embryonic development [71, 72]. DNA methylation markers have been detected in the spermatogonia stage; therefore, the abnormal DNA methylation patterns observed in infertile men may be due to the failure of re-methylation in spermatogonia or alterations to methylation maintenance in spermatocytes, sperm cells or the mature sperm cell. In addition, abnormal DNA methylation may be associated with the abnormal activation of DNMTs [73, 74]. However, earlier studies on candidate genes have shown a strong association between abnormal semen parameters and aberrant DNA methylation in imprinted, testes-specific and other genes [75–77]. Genome-wide methylation analysis by using 450 K BeadChip on spermatozoal DNA from six infertile and six fertile men to identify DMCs showed that the methylation changes in a number of genes have been correlated with reduced sperm count and motility. Also, the loss of spermatogenesis and fertility was correlated with 1680 differentially-methylated CpGs (DMCs) across 1052 genes [35]. The DNA methylation profile of the imprinted genes MEST and SNRPN DMRs have been recently investigated by using meta-analysis. Comparisons of sperm DNA methylation aberrations showed that these genes have higher methylation levels in idiopathic infertile men than in fertile men [78]. Among other differentially-methylatyed genes, CRISPLD1 may have a role in cellular adhesion, which is essential for fertilization [79]. Also, poor semen parameters showed to be most affected by H19 hypomethylation [80].

Moreover, many clinical studies have established that germ cells in adult male mice have a unique genome-wide pattern of DNA methylation. In testicular DNA, the level of DNA methylation is known to be high, which may be associated with the hypermethylation of the CpG rich region as compared with somatic tissues [81, 82]. However, a previous analysis demonstrated that altered expression of spermatogenesis genes was associated with abnormal DNA methylation. Interestingly, poor semen parameters in men were associated with these genes defects [29]. These finding are in line with other results that showed that the oligozoospermic men have different DNA methylation profiles as compared with normozoospermic controls [83]. As a result, the specific sperm DNA methylation in mammals is suggested to be essential for spermatogenesis, fertilization and early embryonic development.

DNA methylation profile of gametes is of a particular importance because it is one of the factors that control the expression of imprinted genes that are crucial for embryonic development, fetal growth, and post-natal behavior [84]. In addition, the results of rodent models suggest that DNA methylation may be involved in the pathogenesis of human male infertility through spermatogenesis alteration. DNA methylation is a reversible epigenetic mark and its effects may be reversed by using either demethylating agents, like DNA methyltransferase (DNMT) inhibitors, 5-azacitidine, and 5-aza-20 –deoxycytidine, or methyl donors, like choline, methionine, and folate [85, 86]. However, a new study suggest that sperm methylation is a possible mechanism of age-induced poor reproductive outcomes among couples undergoing infertility treatment and that it can be used to identify the possible candidate genes for mediating the effects [87]. On the other hand, sperm cells from Norwegian red bulls of inferior fertility have less compact chromatin structure, higher levels of DNA damage, and are hypermethylated compared with bulls of superior fertility [88].

The study of Ni et al. showed that the level of TET1–3 expression is pivotal for male fertility and that TET enzymes are successively expressed at different stages of human spermatogenesis [89]. More recently, it was also reported that modifications in 5hmC pattern in sperms are associated with male infertility. In these reports, infertile males were shown to contain higher rate of 5hmC than fertile males and infertility was correlated with defects in sperm morphology and a high sperm DNA fragmentation rate [90]. Recent results also showed that Tet1-deficient mice undergo a progressive reduction of spermatogonia stem cells and spermatogenesis and thus accelerated infertility with age. Tet1 deficiency decreases 5hmC levels in spermatogonia and downregulates a subset of genes important for cell cycle, germ cell differentiation, meiosis and reproduction, such as Ccna1 and Spo11, resulting in premature reproductive aging [91]. However, during spermatogenesis, 5hmC level is changed dynamically and correlated with gene expression, and RNAseq data shows that Tet1 gene is expressed in spermatogonia [92].

3.2 Disrupted sperm's histone marks

Demethylation of histone plays critical roles in the regulation of gene expression during the spermatogenesis, which may lead to male infertility [22]. In normal human sperm, histone modifications and their enrichment patterns suggest a highly regulated epigenetic landscape. However, it has been reported that aberrant histone methylation and/or acetylation are implicated in the mature sperm in various forms of infertility [17]. Chon and colleagues investigated the histone post-translational modification status in normal and abnormal sperm samples; they found significant histone post-translational modification alterations in abnormal sperm samples compared to those of normal spermatozoa [93]. Sperm histones are involved in chromatin conformation and gene expression through various posttranslational modifications. Modification of sperm histones may influence the large scale demethylation wave after fertilization, altering the expression of offspring genes [68].

Multiple histone variants found in sperms play an essential role throughout spermatogenesis as well as in the mature spermatozoa. Among these, important nuclear proteins are histone 2A and B (H2A and H2B), histone 3 (H3), histone 4 (H4), and the testes variant (tH2B) [94]. The amino terminus of the four core histones (H2A, H2B, H3, and H4) stores a rich source of genetic information. The residues are subjected to enzymatic reactions to produce posttranslational modifications, which transmit epigenetic information transgenerationally. H3 at lysine 4 (K4) is specifically methylated by Set9, which is related to the activation of gene transcription [95]. La Spina et al. [96] investigated the acetylation status of H3K4Ac and H4K5Ac and the methylation profile of H3K4Me3, H3K4Me, H3K9Me2, H3K79Me2 and H3K36Me3 in abnormal and normal human sperm samples. The authors found partial heterogeneous modifications of histones and the existence of H3K4Me1, H3K9Me2, H3K4Me3, H3K79Me2 and H3K36Me3 marks in normal spermatozoa [96].

Several studies have shown that some novel Histone posttranslational modifications (HPTMs) are associated with sperm maturation disorders and dysgenesis. These abnormalities are closely connected with the bromodomain of BRDT, which can recognize histone Kac and recruit transcription complexes from chromatin to promote specific gene expression [97]. However, defects in either the replacement or the modification of histones might result in azoospermia, oligospermia or teratozoospermia, which leads to male infertility [98]. In general, deviations in the sperm histone code have been associated with sperm incompetency and decreased fertility [29]. For example, the role of dimethylation of lysine K4 on histones H3 (H3K4me2) is well-studied in sperm abnormalities. It has been detected at the promoters of transcriptionally active housekeeping genes and indispensable genes that play an important role in the development the spermatogenesis processes [99]. Also, studies in TH2A/TH2B double knockout male mice indicate its role in chromatin compaction and male fertility [94].

3.3 Sperm's chromatin remodeling and non-coding RNA

Chromatin remodeling is the dynamic modification of chromatin architecture to allow access of condensed genomic DNA to the regulatory transcription machinery proteins, and thereby control gene expression. Chromatin remodeling is a landmark event in spermatogenesis, during which transition of nucleohistone to nucleoprotamine takes place in male germ cells. It is initiated by histone hyperacetylation, followed by replacement of somatic histones with testis specific histone variants [98]. However, Sperm chromatin reorganization is an important process that allows spermatozoa to pack huge amounts of DNA into a small nucleus [3]. Patankar et al. [94] revealed that chromatin compaction is positively correlated with sperm- motility, concentration, viability and transcript levels of PRKAG2 and CATSPER B. Also, the authors found that the altered expression of TH2B associated genes in infertile individuals with sperm chromatin compaction defects indicates involvement of TH2B in transcriptional regulation of these genes in post meiotic male germ cells. This altered transcriptome may be either a consequence or a cause of abnormal

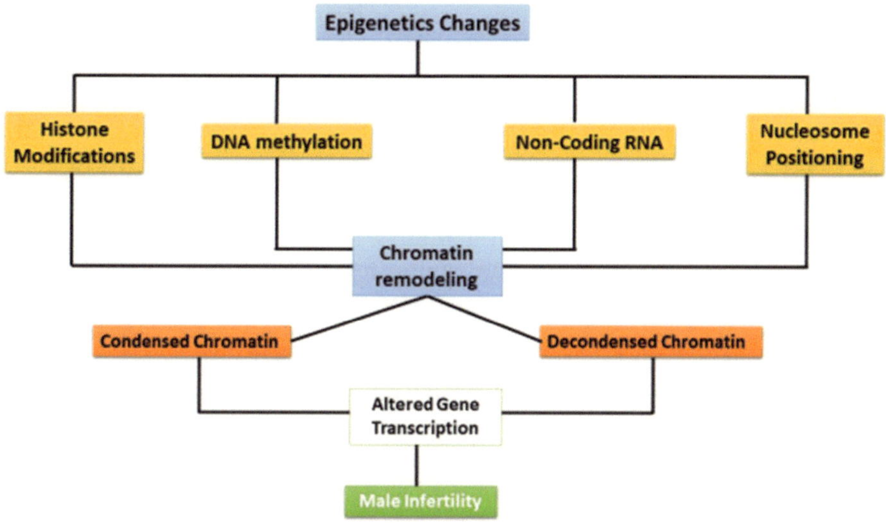

Figure 1.
A schematic diagram summarizing the main epigenetic changes associated with male infertility.

nuclear remodeling during spermatogenesis [94]. Indeed, spermatogenesis offers a unique process to study mechanisms of chromatin remodeling. However, this process is currently understudied and still poorly understood, mainly due to the complexity of the process itself and lack of in vitro experimental systems for studying it [100].

Non-coding RNAs (ncRNAs), as a mark in the epigenetics of germ cells, can be classified into two master groups according to their length as small non-coding and long non-coding RNAs (lncRNAs). However, many potential lncRNAs have been identified in male germ cell development and male infertility, but till date only few have been functionally characterized by gene specific studies [101]. Zhang et al. undertook sequencing to identify lncRNAs that differ in motile and immotile human sperms. While 9879 lncRNA genes (13,819 lncRNA transcripts) showed differential expression between motile and immotile sperms, three lncRNAs, i.e. lnc32058, lnc09522, and lnc98497, showed specific and high expression in immotile sperm in comparison to normal motile sperm [53]. Short ncRNAs can be grouped into three major classes called miRNAs, siRNAs and piRNAs. However, piRNAs are only expressed at the pachytene stage in spermatocytes and in round spermatids. These 30 nucleotide-long piRNAs are involved in sperm maturation and interact with the piwi proteins. Germ line mutations of piwi have been found to prevent piwi ubiquitination and degradation in patients with azoospermia [3, 102]. Understanding the mechanisms underlying non-coding RNA is particularly important in order to develop therapeutic strategies for male genital system diseases caused by abnormal sperm. Studies on the mechanisms underlying the regulation of non-coding RNA during spermatogenesis are still in their initial stages. Numerous issues remain, such as transgenerational inheritance of human epigenetic genes and the association between non-coding RNA and other epigenetic factors (**Figure 1**).

4. Environment and epigenetic in male infertility

The global environment has changed over time as a consequence of industrialization and the progressive accumulation of synthetic pollutants. Such compounds can present in every manufactured product with which humans have contact including cosmetics, food items and containers, packaging materials, toys, agrochemicals. Some of these pollutants take the form of endocrine disruptors that could act with others to alter the ecological balances in natural populations and affected human health and associated with increased incidence of reproductive disease [103–105].

The development and normal functioning of male reproductive system are thought to be highly sensitive to environmental contaminants exposure/insults and metabolic status that could adversely affect sperm's number, quality and the reproductive health of the subjected individuals. In line with the thought that the epigenome is more vulnerable than the genome for such environmental insults, a large number of studies have investigated the role of epigenetic modifications in shaping endocrine functions and their potential influence on spermatogenesis. Due to the protracted period of replication and cell division along the continuous cycles of mitosis and meiosis in adult males spermatogenesis, it is thought that the accumulation of environmentally induced epigenetic are much greater in males than in females [106]. In addition to studies that highlighted dynamical reaction of sperm epigenome to a wide range of environmental and lifestyle stressors [105, 107]. Sperm epigenome is believed to be affected by a large number of biological factors (including aging, obesity, diet, endocrine disruptors and disease), environmental exposures (such as smoking, alcohol, medications, air pollutions, toxic waste socio-economic stress) and life style (i.e. exercise intervention). These factors might contribute to primary sources of the increased male factor infertility and decline in seminal parameters [108]. In comparison to the reproductive system's genome, its cellular epigenetic landscape shows a high degree of plasticity, and thus it is more susceptible to be influenced by the environment insults.

Indeed, the different critical timeframes of spermatozoa development represent windows of susceptibility for epigenetic errors to occur and aberrations potentially induced by environmental insults, possibly affecting fertility and embryonic competence. In an attempt to address the question whether genetic predisposition or environment have significant impact on an individual being infertile, earlier monozygotic twins studies have concluded that socioeconomic environment seemed to influence relative magnitude and pattern with certain genetic background [109]. Based on the findings which emerged from the statistical analysis of heath survey on 1795 Vietnamese male twin cohort, "factors unique" to individual twins could influence to their infertile state more prominently than additive genetic or the common environment effects [110].

Considering both the high rate of unexplained male infertility (up to approximately 50%, [111]) and the lower sensitivity (15%) of semen analysis in predicting infertility [112], the epigenetic disruptions induced by environmental is hoped to contribute in a deeper understanding of male reproductive problems causality and evaluating forms of therapeutic strategies to counteract male infertility. Thus it is believed that environmental epigenetics is considered as the primary molecular actions involve in the increased male infertility and decline in seminal parameters [108]. In both animal model and human studies, a number of defined toxicants and other environmental stimuli exposures have reported to promote testis effects and alter the epigenetic marks that affect spermatogenesis and conferring poor sperm parameters and male infertility. It is expected that sperm differentiation anomalies are attributed to influence of epigenetic aberrations including histone modification and abnormal DNA methylation in imprinted and reproduction associated genes [113].

A large number of studies have shown that exposures to environmental factors, either synthetic or natural origins such as toxicants or nutrition, can have influence on testis biology and male fertility. However, the vast majority of environmental factors is believed to not induce alterations in the DNA sequence but more likely to produce epigenetic alteration. Such environmentally disrupted epigenetic modifications are thought to generate phenotypic variation that includes the induction of disease such as subfertility and imprinting disorders [114]. In this regard, long term exposure of adult rat to butyl-paraben, previously known to affect male rodent reproductive parameters, including testosterone levels and sperm production, resulted in increased DNA methylation changes in the sperm [115]. In respect to the influence on imprinted gene, prenatal exposure to ethanol has also been shown to induce decreased spermatogenesis and loss DNA methylation of the imprinted gene *H19* in mice model studies [116]. Decreased DNA methylation of the *H19* has been suggested to be involved in human male infertility [117, 118]. Another interesting observation relates to exogenous follicle stimulating and luteinizing hormones to immature rats caused epigenetic changes represented by hypomethylation of seminiferous tubular and Leydig cells [119]. Studies have also shown that the epigenetic marks of spermatogenesis are dynamic and can be modulated by micronutrients environment interactions [120]. A number of different environmental toxicants, that take the form of endocrine disruptors, have been shown to promote exposure-specific alterations in the F3 generation sperm epigenome (DNA methylation). These pollutants include the influence of pesticides, insecticide such as DEET [121], hydrocarbon mixture (e.g. jet fuel [122]), dioxin [123] . The findings from aforementioned cited studies suggest that such effects could also induce transgenerational epigenetic phenotypes of inheritance disease and sperm epimutations.

4.1 Epigenetic mechanisms are sensitive to environment and lifestyle: response to antioxidant stress and life style influence as proof of principle

4.1.1 Antioxidant stress

The epigenetic profile of sperm cell has also shown to be influenced by oxidative stress. In this regard, environmental endocrine toxicants such as Bisphenol A are shown to promote male infertility through oxidative DNA damage leading epigenetic modifications in sperm cells mediated by hormonal imbalance [124]. Studies have shown a negative correlation between sperm DNA methylation and sperm DNA fragmentation, as well as for seminal reactive oxygen species (ROS) production. Interestingly, antioxidant supplementation appears to have the potential to reduce DNA damage and normalize sperm DNA methylation in infertile subjects [125]. ROS can lead to chronic inflammation and the latter is closely related to epigenetics as epigenetic aberrations can induce inflammation and vice versa. Various external factors that confer aberrant epigenetic alteration are thought to develop a pro-inflammatory phenotype that contributing to male infertility [126]. Sperm morphological anomalies associated aberrant DNA methylation patterns have been reported experimental rat exposed to uranium [127].

4.2 Life style

A further support to the notion that the main factor driving epigenetic remodeling is induced by external influences came from studies focused on the influence in life style. Modulating of life style factors has shown to influence the sperm's epigenetic landscape by inducing aberrant DNA methylation marks. Of interest, endurance training intervention for relatively short period, 3 months, has resulted in altering the methylation patterns of genes related to the development of the central nervous system, neurogenesis, neuron differentiation and linked to numerous diseases such as schizophrenia and Parkinson's disease [128]. Similar finding was reported by Dokin and colleagues when losing weight induced through gastric-bypass give rise to dynamic changes in the epigenome of human spermatozoa in genes implicated in the central control of appetite under such environmental pressure [129].

In respect to the investigation of the potential influence of nutrition on the sperm epigenome, mice fed low protein diet induced substantial (~30%) increase in methylation at an intergenic CpG island ~50 kb upstream of Ppara [130]. Additionally, prenatal undernutrition can alter DNA methylation in the sperm of adult offspring at regions resistant to zygotic reprogramming such as in long interspersed nuclear elements (LINEs) and short interspersed nuclear elements (SINEs). Studies have concluded that utero undernutrition profoundly disrupt DNA remethylation during germline reprogramming, and contributing to germline hypomethylation [131]. Similarly, high fat diet (HFD) alters sperm cells epigenome of sperm cells reprograming, thereby affecting metabolic tissues of offspring throughout two generations [132]. Through such effect, HFD can modulate spermatogenesis via increasing chromatin compaction and affecting the transcription profile of the spermatozoid [133]. Within this context, HFD-induced obesity provides a link with disease related phenotypes in male infertility, i.e. the impaired quality of spermatozoa from obese men, and transgenerational epigenetic modifications [134]. Interestingly, an altered signature of piRNA, that is known to cause epigenetic changes, was identified in sperms from obese men and rats than their lean counterparts [132]. In line with this evidence, the expression pattern of piRNAs is significantly changed in SPZ when lean and healthy individuals subjected to endurance training highlighting the potential effect of lifestyle on the dynamic expression of sperm [135].

The daily consumption of nutrients and micronutrients represents a source of methyl and acetyl groups that are necessary for the epigenetic dynamics and thus epigenetic modifications targeting spermatogenesis markedly affected by diet depletion of specific nutrients. In this regards, human sperm quality showed to be enhanced by a dietary supplementation with folate [136] and vitamin D [137]. The latter is known to have a key role in the epigenetic routes as genes involved in the vitamin D signaling pathway have several CpG islands in their promoters that can be methylated [138]. Paternal folate-deficient diet is able to induce malformations in the offspring with relevant altered methylation in sperm DNA [139]. Further support to these observations came from studies that linked male infertility onset to the mutations of Mthfr gene, a key enzyme in folate and methionine metabolism [140, 141]. Accordingly, a growing area of research exploring agents able to modulate methyltransferase and histone deacetylase enzyme activities, as nutraceutical agents, has been of an increasing interest.

Considering the dramatic shift in the eating habits, higher calorie intake and the ingestion of highly processed foods and animal products of modern societies, thus, dietary modulation of epigenetic information in spermatogenesis remains an interesting issue currently under investigation with the aim of better understanding the mechanisms underlying transgenerational transmission of environmental conditioning and of evaluating forms of therapeutic strategies to counteract male sub/infertility.

Understanding the role of environmentally driven gametes' epigenetic changes driven by environmental stimuli and their potential effects on the next generations' phenotype could be considered in future risk assessments of male infertility and could add with the developing of important preventing and treating infertility strategies. **Table 1** summaries some of the reported sperm quality anomalies and epigenetic alterations in respect to environmental stimuli exposure.

Chemicals	Use	Influence on male reproductive parameters	Impact on sperm's epigenome	References
Atrazine	Herbicide	Affects meiosis, spermatogenesis and reduces sperm output in mice following in utero exposure	Global decrease of H3K4me3 and transcription dysregulation	[142]
p,p'-DDE	Organochlorine pesticide DDT	Impairs testis histology and male fertility	Igf2 hypomethylation in sperm and hypomethylation of the H19-imprinted gene spermatogonia impairment in prepubertal and pubertal rats.	[143, 144]
Vinclozolin	fungicide	Causes transgenerational sperm epimutations. Effects sperm motility, counts, daily sperm production [145].	Alterations in sperm DNA methylation, non-coding RNA expression, and histone retention	[146, 147]
Carbendazim and chlorothalonil		Impaired spermatogenesis of pubertal mice via estrogen receptor (ER) signaling modulation	Disturbance of the global DNA methylation and histone methylation observed in mice	[148, 149]
Zearalenone	A mycotoxin produced by Fusarium	Acts via the ER signaling pathway to impair mouse spermatogenesis by producing elevated DNA double stranded breaks and decreasing the number of spermatogenic cells	Promote a global decrease in DNA methylation, an increase in the methylation of histone marker H3K27 and a decrease of estrogen alpha in the testis of pubertal CD1 mice exposed to a dose lower	[150]
PCB Arochlor 1254	Industrial compounds used in electrical equipment & building materials	Impaired Sertoli cells by decreasing the expression of both follicle-stimulating hormone receptor (FshR) and androgen receptor (AR);	Increase the protein levels of the enzymes DNMT1, DNMT3ab, DNMT3l leading transcriptional gene repression observed in the Sertoli cells	[151]
Bisphenol A (BPA)	Production of polycarbonate plastics and epoxy resins	Induced spermatogenesis dysfunction and reduced sperm quality.	Alter methylation of the ER alpha promoter and enhance the expression of the enzymes DNMT3a and DNMT3b at both transcript and protein levels in adult rat testis, a global increase of both genome-wide and locus-specific methylation in these spermatocytes.	[151, 152]
DEHP	Plasticizer diethylhexyl phthalate	Significantly disrupt spermatogenesis in mice exposed to DEHP	number of differentially methylated regions across the genome showing differences towards FVB/N mice strain	[153]

Chemicals	Use	Influence on male reproductive parameters	Impact on sperm's epigenome	References
Melatonin	Insomnia and improving sleep medication	Helps maintain a normal spermatogenesis and male fertility	Increase of histone methyltransferase ESET abundance, besides diminishing apoptosis and the global increase of H3K9me3.	[154]
Fat -HFD	A food source	HFD-induced DNA modifications in gametogenesis HFD-modulated spermatogenesis, leading to DNA hyper-compaction in SPZ.	Induce specific histone marks variation in mature sperm, especially through H3K9me3 and H3K27me3.	[130]

Table 1.
Sperm epigenetic alterations upon exposure to some of the studied hazardous environmental substances/lifestyle stressors and their associated influence on male reproductive parameters.

5. Conclusions and future prospective

Over all, studies have revealed the effects of several epigenetic factors on infertility in men, including histone modification, defects in chromatin-modifying complexes, and methylation modification in promoters of various genes. At present, the available treatments do not account for all infertile men, and this is especially important for idiopathic infertility. Regarding the epigenetic role in male infertility, recognizing epigenetic mechanisms enables us to develop new epidrugs that can be used in the treatment of infertility in near future. Understanding the role of environmentally driven epigenetic changes in gametes on the phenotype of the offspring constitutes not only a fascinating biological question on its own but also represents a moral obligation for the health of future generations. Due to this, the main focus of recent epigenetic research would be to focus on discovering new factors involved in altering chromatin state and further looking at its involvement in diseased and normal tissue.

Author details

Hayfa H. Hassani, Rakad M. Kh AL-Jumaily and Fadhel M. Lafta*
Department of Biology, College of Science, University of Baghdad, Baghdad, Iraq

*Address all correspondence to: fadhellafta@sc.uobaghdad.edu.iq

IntechOpen

© 2021 The Author(s). Licensee IntechOpen. This chapter is distributed under the terms of the Creative Commons Attribution License (http://creativecommons.org/licenses/by/3.0), which permits unrestricted use, distribution, and reproduction in any medium, provided the original work is properly cited.

References

[1] Kamiński, P., et al., *External and genetic conditions determining male infertility.* International Journal of Molecular Sciences, 2020. **21**(15): p. 5274.

[2] Hassani, H.H., et al., *AZF Microdeletions in Human Semen Infected with Bacteria.* Online Journal of Health and Allied Sciences, 2011. **10**(3).

[3] Gunes, S. and S.C. Esteves, *Role of genetics and epigenetics in male infertility.* Andrologia, 2021. **53**(1): p. e13586.

[4] Dada, R., et al., *Epigenetics and its role in male infertility.* Journal of assisted reproduction and genetics, 2012. **29**(3): p. 213-223.

[5] Heyn, H., et al., *Epigenetic disruption of the PIWI pathway in human spermatogenic disorders.* PloS one, 2012. **7**(10): p. e47892.

[6] Martinez-Arguelles, D., et al., *In utero exposure to di-(2-ethylhexyl) phthalate decreases mineralocorticoid receptor expression in the adult testis.* Endocrinology, 2009. **150**(12): p. 5575-5585.

[7] Shanker, S., Z. Hu, and M. Wilkinson, *Epigenetic regulation and downstream targets of the Rhox5 homeobox gene.* International journal of andrology, 2008. **31**(5): p. 462-470.

[8] Condorelli, R., A.E. Calogero, and S. La Vignera, *Relationship between testicular volume and conventional or nonconventional sperm parameters.* International Journal of Endocrinology, 2013. **2013**.

[9] Godmann, M., S. Zemter, and C. Kosan, *Genetic and Epigenetic Mouse Models of Human Male Infertility*, in *Genetics of Human Infertility*. 2017, Karger Publishers. p. 143-161.

[10] Luján, S., et al., *Sperm DNA methylation epimutation biomarkers for male infertility and FSH therapeutic responsiveness.* Scientific reports, 2019. **9**(1): p. 1-12.

[11] Wykes, S.M. and S.A. Krawetz, *The structural organization of sperm chromatin.* Journal of Biological Chemistry, 2003. **278**(32): p. 29471-29477.

[12] Fuentes-Mascorro, G., H. Serrano, and A. Rosado, *Sperm chromatin.* Archives of andrology, 2000. **45**(3): p. 215-225.

[13] Oliva, R., *Protamines and male infertility.* Human reproduction update, 2006. **12**(4): p. 417-435.

[14] Said, A.A. and A. Agarwal, *Sperm chromatin assessment.* Textbook of assisted reproductive techniques Laboratory perspectives, 2005. **1**: p. 75.

[15] Engel, W., et al., *The genes for protamine 1 and 2 (PRM1 and PRM2) and transition protein 2 (TNP2) are closely linked in the mammalian genome.* Cytogenetic and Genome Research, 1992. **61**(2): p. 158-159.

[16] Meistrich, M.L., et al., *Roles of transition nuclear proteins in spermiogenesis.* Chromosoma, 2003. **111**(8): p. 483-488.

[17] Jenkins, T.G. and D.T. Carrell, *The sperm epigenome and potential implications for the developing embryo.* Reproduction, 2012. **143**(6): p. 727.

[18] Carrell, D.T. and S.S. Hammoud, *The human sperm epigenome and its potential role in embryonic development.* MHR: Basic science of reproductive medicine, 2009. **16**(1): p. 37-47.

[19] Lachner, M., R.J. O'Sullivan, and T. Jenuwein, *An epigenetic road map for*

histone lysine methylation. Journal of cell science, 2003. **116**(11): p. 2117-2124.

[20] Khalil, A.M., F.Z. Boyar, and D.J. Driscoll, *Dynamic histone modifications mark sex chromosome inactivation and reactivation during mammalian spermatogenesis.* Proceedings of the National Academy of Sciences, 2004. **101**(47): p. 16583-16587.

[21] Payne, C. and R.E. Braun, *Histone lysine trimethylation exhibits a distinct perinuclear distribution in Plzf-expressing spermatogonia.* Developmental biology, 2006. **293**(2): p. 461-472.

[22] Grimes Jr, S.R. and N. Henderson, *Hyperacetylation of histone H4 in rat testis spermatids.* Experimental cell research, 1984. **152**(1): p. 91-97.

[23] Jenuwein, T. and C.D. Allis, *Translating the histone code.* Science, 2001. **293**(5532): p. 1074-1080.

[24] Berger, S.L., Histone modifications in transcriptional regulation. Current opinion in genetics & development, 2002. 12(2): p. 142-148

[25] Fernandez-Capetillo, O., et al., *H2AX is required for chromatin remodeling and inactivation of sex chromosomes in male mouse meiosis.* Developmental cell, 2003. **4**(4): p. 497-508.

[26] Zhu, B., et al., *Monoubiquitination of human histone H2B: the factors involved and their roles in HOX gene regulation.* Molecular cell, 2005. **20**(4): p. 601-611.

[27] Erkek, S., et al., *Molecular determinants of nucleosome retention at CpG-rich sequences in mouse spermatozoa.* Nature structural & molecular biology, 2013. **20**(7): p. 868.

[28] Rathke, C., et al., *Chromatin dynamics during spermiogenesis.* Biochimica et Biophysica Acta (BBA)-Gene Regulatory Mechanisms, 2014. **1839**(3): p. 155-168.

[29] Cui, X., et al., *DNA methylation in spermatogenesis and male infertility.* Experimental and therapeutic medicine, 2016. **12**(4): p. 1973-1979.

[30] Curradi, M., et al., *Molecular mechanisms of gene silencing mediated by DNA methylation.* Molecular and cellular biology, 2002. **22**(9): p. 3157-3173.

[31] He, Y.-F., et al., *Tet-mediated formation of 5-carboxylcytosine and its excision by TDG in mammalian DNA.* Science, 2011. **333**(6047): p. 1303-1307.

[32] Ito, S., et al., *Tet proteins can convert 5-methylcytosine to 5-formylcytosine and 5-carboxylcytosine.* Science, 2011. **333**(6047): p. 1300-1303.

[33] Biermann, K. and K. Steger, *Epigenetics in male germ cells.* Journal of andrology, 2007. **28**(4): p. 466-480.

[34] Tang, W.W., et al., *Specification and epigenetic programming of the human germ line.* Nature Reviews Genetics, 2016. **17**(10): p. 585.

[35] Sujit, K.M., et al., *Genome-wide differential methylation analyses identifies methylation signatures of male infertility.* Human Reproduction, 2018. **33**(12): p. 2256-2267.

[36] Omisanjo, O.A., et al., *DNMT1 and HDAC1 gene expression in impaired spermatogenesis and testicular cancer.* Histochemistry and cell biology, 2007. **127**(2): p. 175-181.

[37] Rousseaux, S., et al., *Épigénétique du spermatozoïde.* Gynécologie obstétrique & fertilité, 2006. **34**(9): p. 831-835.

[38] La Salle, S. and J.M. Trasler, *Dynamic expression of DNMT3a and DNMT3b isoforms during male germ cell development in the mouse.* Developmental biology, 2006. **296**(1): p. 71-82.

[39] Marques, C.J., et al., *DNA methylation imprinting marks and DNA methyltransferase expression in human spermatogenic cell stages.* Epigenetics, 2011. **6**(11): p. 1354-1361.

[40] Riesewijk, A.M., et al., *Monoallelic expression of humanPEG1/MESTIs paralleled by parent-specific methylation in fetuses.* Genomics, 1997. **42**(2): p. 236-244.

[41] Kerjean, A., et al., *Establishment of the paternal methylation imprint of the human H19 and MEST/PEG1 genes during spermatogenesis.* Human Molecular Genetics, 2000. **9**(14): p. 2183-2187.

[42] Li, Z., et al., *Distinct roles of DNMT1-dependent and DNMT1-independent methylation patterns in the genome of mouse embryonic stem cells.* Genome biology, 2015. **16**(1): p. 1-15.

[43] Donkin, I. and R. Barrès, *Sperm epigenetics and influence of environmental factors.* Molecular metabolism, 2018. **14**: p. 1-11.

[44] Matsuoka, T., et al., *DNA methyltransferase-3 like protein expression in various histological types of testicular germ cell tumor.* Japanese journal of clinical oncology, 2016. **46**(5): p. 475-481.

[45] Mäkelä, J.-A., et al., *Testis development.* Endocrine reviews, 2019. **40**(4): p. 857-905.

[46] Meikar, O., et al., *Chromatoid body and small RNAs in male germ cells.* Reproduction, 2011. **142**(2): p. 195-209.

[47] Yu, Z., T. Raabe, and N.B. Hecht, *MicroRNA Mirn122a reduces expression of the posttranscriptionally regulated germ cell transition protein 2 (Tnp2) messenger RNA (mRNA) by mRNA cleavage.* Biology of reproduction, 2005. **73**(3): p. 427-433.

[48] Johnson, G.D., et al., *The sperm nucleus: chromatin, RNA and the nuclear matrix.* Reproduction (Cambridge, England), 2011. **141**(1): p. 21.

[49] Watanabe, T., et al., *Identification and characterization of two novel classes of small RNAs in the mouse germline: retrotransposon-derived siRNAs in oocytes and germline small RNAs in testes.* Genes & development, 2006. **20**(13): p. 1732-1743.

[50] Aravin, A.A., et al., *A piRNA pathway primed by individual transposons is linked to de novo DNA methylation in mice.* Molecular cell, 2008. **31**(6): p. 785-799.

[51] Reuter, M., et al., *Miwi catalysis is required for piRNA amplification-independent LINE1 transposon silencing.* Nature, 2011. **480**(7376): p. 264-267.

[52] Bao, J., et al., *Expression profiling reveals developmentally regulated lncRNA repertoire in the mouse male germline.* Biology of reproduction, 2013. **89**(5): p. 107, 1-12.

[53] Zhang, X., et al., *Expression profiles and characteristics of human lncRNA in normal and asthenozoospermia sperm.* Biology of reproduction, 2019. **100**(4): p. 982-993.

[54] Rathke, C., et al., *Distinct functions of Mst77F and protamines in nuclear shaping and chromatin condensation during Drosophila spermiogenesis.* European journal of cell biology, 2010. **89**(4): p. 326-338.

[55] Ørom, U.A. and R. Shiekhattar, *Long noncoding RNAs usher in a new era in the biology of enhancers.* Cell, 2013. **154**(6): p. 1190-1193.

[56] Ji, P., et al., *N6-Methyladenosine in RNA and DNA: an epitranscriptomic and epigenetic player implicated in determination of stem cell fate.* Stem cells international, 2018. **2018**.

[57] Zhang, M., et al., *Roles of N6-Methyladenosine (m6A) in stem cell fate decisions and early embryonic development in mammals.* Frontiers in Cell and Developmental Biology, 2020. **8**: p. 782.

[58] Niu, Y., et al., *N6-methyl-adenosine (m6A) in RNA: an old modification with a novel epigenetic function.* Genomics, proteomics & bioinformatics, 2013. **11**(1): p. 8-17.

[59] Tian, S., et al., *Regulation of gene expression associated with the N6-methyladenosine (m6A) enzyme system and its significance in cancer.* Frontiers in Oncology, 2020. **10**: p. 3123.

[60] Liu, S., et al., *Role of RNA N6-Methyladenosine Modification in Male Infertility and Genital System Tumors.* Frontiers in Cell and Developmental Biology, 2021. **9**.

[61] Lasman, L., J.H. Hanna, and N. Novershtern, *Role of m6A in embryonic stem cell differentiation and in gametogenesis.* Epigenomes, 2020. **4**(1): p. 5.

[62] Xu, K., et al., *Mettl3-mediated m 6 A regulates spermatogonial differentiation and meiosis initiation.* Cell research, 2017. **27**(9): p. 1100-1114.

[63] Lin, Z., et al., *Mettl3-/Mettl14-mediated mRNA N 6-methyladenosine modulates murine spermatogenesis.* Cell research, 2017. **27**(10): p. 1216-1230.

[64] Zheng, G., et al., *ALKBH5 is a mammalian RNA demethylase that impacts RNA metabolism and mouse fertility.* Molecular cell, 2013. **49**(1): p. 18-29.

[65] Zhao, T., et al., *Increased m6A modification of RNA methylation related to the inhibition of demethylase FTO contributes to MEHP-induced Leydig cell injury☆.* Environmental Pollution, 2021. **268**: p. 115627.

[66] Hsu, P.J., et al., *Ythdc2 is an N 6-methyladenosine binding protein that regulates mammalian spermatogenesis.* Cell research, 2017. **27**(9): p. 1115-1127.

[67] Jain, D., et al., *ketu mutant mice uncover an essential meiotic function for the ancient RNA helicase YTHDC2.* Elife, 2018. **7**: p. e30919.

[68] Xu, X., et al., *Epigenetic Mechanisms of Paternal Stress in Offspring Development and Diseases.* International Journal of Genomics, 2021. **2021**.

[69] Schütte, B., et al., *Broad DNA methylation changes of spermatogenesis, inflammation and immune response-related genes in a subgroup of sperm samples for assisted reproduction.* Andrology, 2013. **1**(6): p. 822-829.

[70] Calicchio, R., et al., *DNA methylation, an epigenetic mode of gene expression regulation in reproductive science.* Current pharmaceutical design, 2014. **20**(11): p. 1726-1750.

[71] Godmann, M., R. Lambrot, and S. Kimmins, *The dynamic epigenetic program in male germ cells: Its role in spermatogenesis, testis cancer, and its response to the environment.* Microscopy research and technique, 2009. **72**(8): p. 603-619.

[72] Oakes, C., et al., *Developmental acquisition of genome-wide DNA methylation occurs prior to meiosis in male germ cells.* Developmental biology, 2007. **307**(2): p. 368-379.

[73] Wasson, J.A., C.C. Ruppersburg, and D.J. Katz, *Restoring totipotency through epigenetic reprogramming.* Briefings in functional genomics, 2013. **12**(2): p. 118-128.

[74] Skinner, M.K., *Environmental epigenomics and disease susceptibility.* EMBO reports, 2011. **12**(7): p. 620-622.

[75] Houshdaran, S., et al., *Widespread epigenetic abnormalities suggest a broad*

DNA methylation erasure defect in abnormal human sperm. PloS one, 2007. **2**(12): p. e1289.

[76] Kobayashi, H., et al., *Aberrant DNA methylation of imprinted loci in sperm from oligospermic patients.* Human molecular genetics, 2007. **16**(21): p. 2542-2551.

[77] Tian, M., et al., *Association of DNA methylation and mitochondrial DNA copy number with human semen quality.* Biology of reproduction, 2014. **91**(4): p. 101, 1-8.

[78] Santi, D., et al., *Impairment of sperm DNA methylation in male infertility: a meta-analytic study.* Andrology, 2017. **5**(4): p. 695-703.

[79] Gibbs, G.M., K. Roelants, and M.K. O'bryan, *The CAP superfamily: cysteine-rich secretory proteins, antigen 5, and pathogenesis-related 1 proteins—roles in reproduction, cancer, and immune defense.* Endocrine reviews, 2008. **29**(7): p. 865-897.

[80] Darbandi, M., et al., *Reactive oxygen species-induced alterations in H19-Igf2 methylation patterns, seminal plasma metabolites, and semen quality.* Journal of assisted reproduction and genetics, 2019. **36**(2): p. 241-253.

[81] Montjean, D., et al., *Sperm transcriptome profiling in oligozoospermia.* Journal of assisted reproduction and genetics, 2012. **29**(1): p. 3-10.

[82] Montjean, D., et al., *Methylation changes in mature sperm deoxyribonucleic acid from oligozoospermic men: assessment of genetic variants and assisted reproductive technology outcome.* Fertility and sterility, 2013. **100**(5): p. 1241-1247. e2.

[83] Montjean, D., et al., *Sperm global DNA methylation level: association with semen parameters and genome integrity.* Andrology, 2015. **3**(2): p. 235-240.

[84] Reik, W. and J. Walter, *Genomic imprinting: parental influence on the genome.* Nature Reviews Genetics, 2001. **2**(1): p. 21-32.

[85] Egger, G., *liang G, aparicio a, Jones Pa.* Epigenetics in human disease and prospects for epigenetic therapy. Nature, 2004. **429**: p. 457-463.

[86] Fowler, B. *Homocysteine: overview of biochemistry, molecular biology, and role in disease processes.* in *Seminars in vascular medicine*. 2005. Copyright© 2005 by Thieme Medical Publishers, Inc., 333 Seventh Avenue, New ....

[87] Oluwayiose, O.A., et al., *Sperm DNA methylation mediates the association of male age on reproductive outcomes among couples undergoing infertility treatment.* Scientific Reports, 2021. **11**(1): p. 1-14.

[88] Narud, B., et al., *Sperm chromatin integrity and DNA methylation in Norwegian Red bulls of contrasting fertility.* Molecular Reproduction and Development, 2021. **88**(3): p. 187-200.

[89] Ni, K., et al., *TET enzymes are successively expressed during human spermatogenesis and their expression level is pivotal for male fertility.* Human Reproduction, 2016. **31**(7): p. 1411-1424.

[90] Efimova, O.A., et al., *Genome-wide 5-hydroxymethylcytosine patterns in human spermatogenesis are associated with semen quality.* Oncotarget, 2017. **8**(51): p. 88294.

[91] Huang, G., et al., *Tet1 deficiency leads to premature reproductive aging by reducing spermatogonia stem cells and germ cell differentiation.* Iscience, 2020. **23**(3): p. 100908.

[92] Hammoud, S.S., et al., *Transcription and imprinting dynamics in developing postnatal male germline stem cells.* Genes & development, 2015. **29**(21): p. 2312-2324.

[93] Schon, S.B., et al., *Histone modification signatures in human sperm distinguish clinical abnormalities.* Journal of assisted reproduction and genetics, 2019. **36**(2): p. 267-275.

[94] Patankar, A., et al., *Epigenetic landscape of testis specific histone H2B variant and its influence on sperm function.* Clinical epigenetics, 2021. **13**(1): p. 1-18.

[95] Nishioka, K., et al., *Set9, a novel histone H3 methyltransferase that facilitates transcription by precluding histone tail modifications required for heterochromatin formation.* Genes & development, 2002. **16**(4): p. 479-489.

[96] La Spina, F.A., et al., *Heterogeneous distribution of histone methylation in mature human sperm.* Journal of assisted reproduction and genetics, 2014. **31**(1): p. 45-49.

[97] Bhattacharya, S., S. Piya, and G. Borthakur, *Bromodomain inhibitors: what does the future hold.* Clin. Adv. Hematol. Oncol, 2018. **16**: p. 504-515.

[98] Wang, T., et al., *Essential role of histone replacement and modifications in male fertility.* Frontiers in genetics, 2019. **10**: p. 962.

[99] Štiavnická, M., et al., *H3K4me2 accompanies chromatin immaturity in human spermatozoa: an epigenetic marker for sperm quality assessment.* Systems biology in reproductive medicine, 2020. **66**(1): p. 3-11.

[100] Li, W., et al., *Chd5 orchestrates chromatin remodelling during sperm development.* Nature communications, 2014. **5**(1): p. 1-15.

[101] Joshi, M. and S. Rajender, *Long non-coding RNAs (lncRNAs) in spermatogenesis and male infertility.* Reproductive Biology and Endocrinology, 2020. **18**(1): p. 1-18.

[102] Song, R., et al., *Male germ cells express abundant endogenous siRNAs.* Proceedings of the National Academy of Sciences, 2011. **108**(32): p. 13159-13164.

[103] Balabanič, D., M. Rupnik, and A.K. Klemenčič, *Negative impact of endocrine-disrupting compounds on human reproductive health.* Reproduction, Fertility and Development, 2011. **23**(3): p. 403-416.

[104] Maqbool, F., et al., *Review of endocrine disorders associated with environmental toxicants and possible involved mechanisms.* Life sciences, 2016. **145**: p. 265-273.

[105] Leisegang, K. and S. Dutta, *Do lifestyle practices impede male fertility?* Andrologia, 2021. **53**(1): p. e13595.

[106] Messerschmidt, D.M., B.B. Knowles, and D. Solter, *DNA methylation dynamics during epigenetic reprogramming in the germline and preimplantation embryos.* Genes & development, 2014. **28**(8): p. 812-828.

[107] Machen, G.L. and J.I. Sandlow, *Causes of Male Infertility*, in *Male Infertility*. 2020, Springer. p. 3-14.

[108] Lujan, S., et al., *Sperm DNA Methylation Epimutation Biomarkers for Male Infertility and FSH Therapeutic Responsiveness.* Sci Rep, 2019. **9**(1): p. 16786.

[109] Kohler, H.P., J.L. Rodgers, and K. Christensen, *Is fertility behavior in our genes? Findings from a Danish twin study.* Population and development review, 1999. **25**(2): p. 253-288.

[110] Cloonan, Y.K., V.L. Holt, and J. Goldberg, *Male factor infertility: a twin study.* Paediatric and perinatal epidemiology, 2007. **21**(3): p. 229-234.

[111] Krausz, C., *Male infertility: pathogenesis and clinical diagnosis.* Best practice & research Clinical

endocrinology & metabolism, 2011. **25**(2): p. 271-285.

[112] Khatun, A., M.S. Rahman, and M.G. Pang, *Clinical assessment of the male fertility*. Obstet Gynecol Sci, 2018. **61**(2): p. 179-191.

[113] Gunes, S., et al., *The role of epigenetics in idiopathic male infertility*. Journal of assisted reproduction and genetics, 2016. **33**(5): p. 553-569.

[114] Inbar-Feigenberg, M., et al., *Basic concepts of epigenetics*. Fertil Steril, 2013. **99**(3): p. 607-615.

[115] Park, C.J., et al., *Butyl paraben-induced changes in DNA methylation in rat epididymal spermatozoa*. Andrologia, 2012. **44 Suppl 1**: p. 187-193.

[116] Stouder, C., E. Somm, and A. Paoloni-Giacobino, *Prenatal exposure to ethanol: a specific effect on the H19 gene in sperm*. Reproductive Toxicology, 2011. **31**(4): p. 507-512.

[117] Li, B., et al., *Altered DNA methylation patterns of the H19 differentially methylated region and the DAZL gene promoter are associated with defective human sperm*. PloS one, 2013. **8**(8): p. e71215.

[118] Nasri, F., et al., *Sperm DNA methylation of H19 imprinted gene and male infertility*. Andrologia, 2017. **49**(10): p. e12766.

[119] Reddy, P.S. and P. Reddy, *Differential regulation of DNS methylation in rat testis and its regulation by gonadotropic hormones*. Journal of steroid biochemistry, 1990. **35**(2): p. 173-178.

[120] Bodden, C., A.J. Hannan, and A.C. Reichelt, *Diet-induced modification of the sperm epigenome programs metabolism and behavior*. Trends in Endocrinology & Metabolism, 2020. **31**(2): p. 131-149.

[121] Manikkam, M., et al., *Pesticide and insect repellent mixture (permethrin and DEET) induces epigenetic transgenerational inheritance of disease and sperm epimutations*. Reproductive toxicology, 2012. **34**(4): p. 708-719.

[122] Tracey, R., et al., *Hydrocarbons (jet fuel JP-8) induce epigenetic transgenerational inheritance of obesity, reproductive disease and sperm epimutations*. Reproductive toxicology, 2013. **36**: p. 104-116.

[123] Manikkam, M., et al., *Dioxin (TCDD) induces epigenetic transgenerational inheritance of adult onset disease and sperm epimutations*. 2012.

[124] Cariati, F., et al., *Bisphenol A-Induced Epigenetic Changes and Its Effects on the Male Reproductive System*. Frontiers in Endocrinology, 2020. **11**.

[125] Tunc, O. and K. Tremellen, *Oxidative DNA damage impairs global sperm DNA methylation in infertile men*. J Assist Reprod Genet, 2009. **26**(9-10): p. 537-544.

[126] Loveland, K.L., et al., *Cytokines in male fertility and reproductive pathologies: immunoregulation and beyond*. Frontiers in Endocrinology, 2017. **8**: p. 307.

[127] Legendre, A., et al., *Multigenerational exposure to uranium changes morphometric parameters and global DNA methylation in rat sperm*. Comptes rendus biologies, 2019. **342** (5-6): p. 175-185.

[128] Denham, J., et al., *Genome-wide sperm DNA methylation changes after 3 months of exercise training in humans*. Epigenomics, 2015. 7(5): p. 717-731.

[129] Donkin, I., et al., *Obesity and bariatric surgery drive epigenetic variation of spermatozoa in humans*. Cell metabolism, 2016. **23**(2): p. 369-378.

[130] Carone, B.R., et al., *Paternally induced transgenerational environmental reprogramming of metabolic gene expression in mammals.* Cell, 2010. **143**(7): p. 1084-1096.

[131] Radford, E.J., et al., *In utero undernourishment perturbs the adult sperm methylome and intergenerational metabolism.* Science, 2014. **345**(6198).

[132] de Castro Barbosa, T., et al., *High-fat diet reprograms the epigenome of rat spermatozoa and transgenerationally affects metabolism of the offspring.* Molecular metabolism, 2016. **5**(3): p. 184-197.

[133] Blin, G., et al., *Maternal exposure to high-fat diet induces long-term derepressive chromatin marks in the heart.* Nutrients, 2020. **12**(1): p. 181.

[134] Houfflyn, S., C. Matthys, and A. Soubry, *Male obesity: epigenetic origin and effects in sperm and offspring.* Current molecular biology reports, 2017. **3**(4): p. 288-296.

[135] Ingerslev, L.R., et al., *Endurance training remodels sperm-borne small RNA expression and methylation at neurological gene hotspots.* Clinical epigenetics, 2018. **10**(1): p. 1-11.

[136] Mendiola, J., et al., *A low intake of antioxidant nutrients is associated with poor semen quality in patients attending fertility clinics.* Fertility and sterility, 2010. **93**(4): p. 1128-1133.

[137] Wadhwa, L., et al., *Impact of Vitamin D Supplementation on Semen Quality in Vitamin D-Deficient Infertile Males with Oligoasthenozoospermia.* The Journal of Obstetrics and Gynecology of India, 2020. **70**(1): p. 44-49.

[138] Pike, J.W., M.B. Meyer, and K.A. Bishop, *Regulation of target gene expression by the vitamin D receptor-an update on mechanisms.* Reviews in Endocrine and Metabolic Disorders, 2012. **13**(1): p. 45-55.

[139] Lambrot, R., et al., *Low paternal dietary folate alters the mouse sperm epigenome and is associated with negative pregnancy outcomes.* Nature communications, 2013. **4**(1): p. 1-13.

[140] Pourborhanzadeh, A., et al., *Association study of MTHFR-c677t with male infertility and reporting new potential SNPS/sequence variants as a sourse of population genetic markers.* Genetika, 2020. **52**(3): p. 1181-1192.

[141] Kos, B.J., et al., *The association of parental methylenetetrahydrofolate reductase polymorphisms (MTHFR 677C> T and 1298A> C) and fetal loss: a case–control study in South Australia.* The Journal of Maternal-Fetal & Neonatal Medicine, 2020. **33**(5): p. 752-757.

[142] Hao, C., et al., *Exposure to the widely used herbicide atrazine results in deregulation of global tissue-specific RNA transcription in the third generation and is associated with a global decrease of histone trimethylation in mice.* Nucleic Acids Res, 2016. **44**(20): p. 9784-9802.

[143] Shi, Y.Q., et al., *p, p'-DDE induces apoptosis and mRNA expression of apoptosis-associated genes in testes of pubertal rats.* Environmental toxicology, 2013. **28**(1): p. 31-41.

[144] Guerrero-Bosagna, C., et al., *Environmentally induced epigenetic transgenerational inheritance of altered Sertoli cell transcriptome and epigenome: molecular etiology of male infertility.* PloS one, 2013. **8**(3): p. e59922.

[145] Feijo, M., et al., *Effects of the endocrine disruptor vinclozolin in male reproduction: a systematic review and meta-analysis.* Biol Reprod, 2021.

[146] Ben Maamar, M., et al., *Alterations in sperm DNA methylation, non-coding RNA expression, and histone retention mediate vinclozolin-induced epigenetic transgenerational inheritance of disease.*

Environmental Epigenetics, 2018. **4**(2): p. dvy010.

[147] Skinner, M.K., et al., *Transgenerational sperm DNA methylation epimutation developmental origins following ancestral vinclozolin exposure.* Epigenetics, 2019. **14**(7): p. 721-739.

[148] Zhang, P., et al., *Low dose chlorothalonil impairs mouse spermatogenesis through the intertwining of estrogen receptor pathways with histone and DNA methylation.* Chemosphere, 2019. **230**: p. 384-395.

[149] Liu, J., et al., *Low dose carbendazim disrupts mouse spermatogenesis might be through estrogen receptor related histone and DNA methylation.* Ecotoxicology and environmental safety, 2019. **176**: p. 242-249.

[150] Pang, J., et al., *Effect of low-dose zearalenone exposure on reproductive capacity of male mice.* Toxicology and applied pharmacology, 2017. **333**: p. 60-67.

[151] Priya, E.S., et al., *Impact of Lactational Exposure to Polychlorinated Biphenyl Causes Epigenetic Modification and Impairs Sertoli Cells Functional Regulators in F1 Progeny.* Reprod Sci, 2018. **25**(6): p. 818-829.

[152] Yin, L., et al., *Role of DNA methylation in bisphenol A exposed mouse spermatocyte.* Environmental toxicology and pharmacology, 2016. **48**: p. 265-271.

[153] Prados, J., et al., *Prenatal exposure to DEHP affects spermatogenesis and sperm DNA methylation in a strain-dependent manner.* PloS one, 2015. **10**(8): p. e0132136.

[154] Lv, Y., et al., *Melatonin protects mouse spermatogonial stem cells against hexavalent chromium-induced apoptosis and epigenetic histone modification.* Toxicology and applied pharmacology, 2018. **340**: p. 30-38.

Chapter 10

The Role of miR-107 in Prostate Cancer: A Review and Experimental Evidence

Maria Elizbeth Alvarez-Sanchez, Oscar Rojas Espinosa, Julio César Torres-Romero, Ereth Ameyatzin Robles Chávez, Edgar Estrella-Parra, María Dolores Ponce Regalado, Raúl Aragón Franco, Jose Gadú Campos Saucedo, Cesar López-Camarillo, Minerva Camacho Nuez and Verónica Fernández Sánchez

Abstract

Over the past two decades, several research groups have focused on the functioning of microRNAs (miRNAs), because many of them function as positive or negative endogenous regulators of processes that alter during the development of cancer. Prostate cancer is the second most commonly occurring cancer in men. New biomarkers are needed to support the diagnosis of prostate cancer. Although it is necessary to deepen the research on this molecule to explore its potential utility in the diagnosis, follow-up, and prognosis of cancer, our results support a role of miR-107 in the signaling cascades that allow cancer progression, and as shown here, in the progression of Prostate Cancer (PCa). These findings strongly suggest that miR-107 may be a potential circulating biomarker for the diagnosis and prognosis of prostate cancer.

Keywords: microRNAs, miR-107, biomarker, cancer hallmarks, therapeutic target

1. Introduction

Cancer can be defined as a state of uncontrolled cell-growth and dissemination that alters several cellular processes and functions [1]. Hanahan and Weinburg described, in 2000, six characteristics of cancer called "cancer hallmarks" that increased to ten in 2011 due to the complexity of the disease and the number of biological mechanisms that become altered. These cancer hallmarks include sustained proliferative signals, evading growth suppressors, avoidance of immune destruction, replicative immortality, tumor-promoting inflammation, sustained invasion and metastasis, induction of angiogenesis, genetic instability and mutation, resistance to cell death, and cell-metabolism deregulation [2].

Currently, more than 100 different types of cancer have been identified. According to the International Agency for Research in Cancer (IARC) in 2020, the most common types of cancer were those of lung, breast, prostate, colon, and stomach [3]. In the 2014 World Cancer Report, the World Health Organization (WHO) reported 14 million new cases and 8.2 million deaths in the year 2012 [4]. Last year, the Pan American Health Organization (PAHO) reported that in the Americas, the most frequently diagnosed types of cancer in men are prostate (21.7%), lung (9.5%), and colorectal (8.0%) cancer [5].

In 2018, the estimated number of new cases increased to 18 million, and it was expected that by 2030, more than 20 million cases will be registered annually [4]. Under this scenario, different methods of early diagnosis and treatment have been sought, which can effectively defeat the disease, leading to a significant reduction in the number of cases.

2. MicroRNAs as a potencial cancer biomarkers

MicroRNAs (miRNAs) are endogenous regulators of different biological processes, including those that are altered in cancer development such as cell growth and proliferation, differentiation, apoptosis, angiogenesis, and others [6]. miRNAs are a family of small non-coding RNAs, 18 to 22 nucleotides long whose function is the post-transcriptional regulation of gene expression [7]. They have a specific region composed of around 6 nucleotides called the seed region that binds the 3'UTR region of the target messenger RNA (mRNA). This union alters the stability of the transcript leading to its degradation or storage in intracellular structures called p-bodies, thus leading to effective repression of translation [8].

Different authors have reported that miRNAs have a tissue-specific expression pattern and that this pattern is altered in cancer tissues [9, 10]. Therefore, it has been suggested that these changes in the expression of miRNAs can be used as possible biomarkers of the disease [11]. However, the standard determination of the relative expression of miRNA requires a tissue sample from the cancer patient, which is considered a highly invasive process [12, 13]. Consequently, other diagnostic methods are required that are safe, effective, and accessible.

In 2008, Mitchell et al. described a specific type of miRNAs that were present in their stable form in different biological fluids, which they called circulating miRNAs (c-miRNAs) [14]. In their research, they analyzed the expression of different miRNAs in plasma, serum, and epithelial tissue samples. They observed that the miRNAs remained stable in serum and plasma, and in the tissue samples. They concluded that blood was a reliable source for the extraction and quantification of miRNAs. These findings have allowed the analysis of the expression of circulating miRNAs as a possible method for cancer diagnosis [14, 15].

There are more than 1000 miRNAs registered in the miRNA database (miRDB) that have been found in the human body; one of them is miR-107, which, according to miRDB, can have more than 800 possible targets [16].

3. Is miR-107, a circulating miRNA?

MicroRNA-107 (miR-107) is a molecule composed of 23 nucleotides; it is considered a c-miRNA because it can be found in a stable form in plasma and urine [17]. This miRNA is highly conserved in humans and other species. In humans, it is transcribed from the introns of the pantothenate kinase 1 gene, located on the long arm of chromosome 10 [16]. It belongs to the miR-103/107 family, which participates in the regulation of proteins involved in cell proliferation, cell cycle arrest, and programmed cell death or apoptosis [18].

Different research groups have reported that miR-107 is altered in different types of cancer in both men and women, including cervical [19], breast [20], ovary [21], colorectal [22], gastric [23], oral [24], penile [25], and prostate [13] cancers. Because of this, it has been proposed as a possible biomarker of cancer and a potential target for treatment.

So far, information on the expression of miR-107 in plasma has been published only on gastrointestinal cancer by Parvaee et al. [15], and in the urine of patients with prostate cancer by Lekchnov et al. [17]. Therefore, there is a great opportunity in this field to contribute to the design of new diagnostic methods for the detection of cancer from the early stages of the disease.

4. miR-107 and its participation in prostate cancer

PCa is the second leading cause of death in men over 45 years of age; the American Cancer Society estimates that 1 in 41 men will die this year from this cause. Nevertheless, prostate cancer has not been studied to the same extent as breast cancer, and hence, there is a lack of information on the development, evolution, and the treatment of this disease. Although the evaluation of prostatic specific antigen (PSA) has contributed to the early detection of PCa, its use may also lead to non-conclusive results because of false positive and negative results. Also, its low specificity can lead to misdiagnosis and incorrect treatment of indolent PCa patients [25, 26]. Patients with prostate cancer who receive radical treatment for presumably locally confined cancer can experience clinical relapses, indicating that the extent of these cancers was greater than the one previously diagnosed [27]. The median survival of metastatic PCa patients is 5 to 8 months, and their 5-year survival rate is less than 30%. This poor prognosis is the result of many factors including the lack of initial symptoms when PCa develops, local invasiveness, or metastases to distant organs in the early stage of the disease, and misdiagnosis [28, 29].

A majority of PCa patients who initially respond to androgen suppression treatment (AST) develop metastatic castration-resistant prostate cancer (CRPCa) within 2 years of treatment [29]. This condition still cannot be predicted, although several biomarkers have been tested (proteomic and metabolomic approaches) [30, 31]. This is mainly due to the wide variability of proteins being tested, the masking effect of abundant serum proteins, the high salt concentration in the samples, and the variability between individuals that drastically reduce the reproducibility of the biomarker's determinations. miRNAs have been found to participate in the alteration of many cellular processes that lead to the development of cancer including proliferation [32, 33], differentiation [34, 35], angiogenesis, and evasion of apoptosis [36]. miRNAs can serve as biomarkers because they are resistant to degradation by RNases either in tumors or serum. miRNAs are small (18-24 nt), endogenous, non-coding RNAs, encoded either on intergenic regions of DNA or on introns and exons of genes that act as post-transcriptional control genes, triggering degradation or blocking translation of mRNAs by complementary base pairing [37]. When discovered, miRNAs were shown to control fundamental cellular processes, such as cell differentiation and timing of the organism development [38, 39], suggesting that aberrations of miRNAs could be involved in various human diseases, including cancer.

In recent years, several detailed studies have described the mechanisms through which miRNAs are stabilized and how they are detected in plasma and serum [10, 14]. Plasma/serum miRNAs are resistant to endogenous ribonuclease activity by binding to specific plasma proteins or by packing into various serum secretory vesicles, including apoptotic bodies and exosomes [14, 40–42]. Various blood-based miRNAs have been identified, including those in this study, and can be used for cancer detection, monitoring tumor dynamics, and predicting prognosis and chemoresistance [43–47].

Several reports have shown that changes in the level of circulating miRNAs associate with prostate cancer [48]. One of these miRNAs is miR-107, which is overexpressed by more than 11-fold in PCa [48] but whose role has not been studied in the context of cancer progression [49].

In 2019, Zhang et al. studied the role of miR-107 in prostate cancer in both the tissue and cancer cells (DU145 and PC3); they found that its levels decreased compared to the levels in healthy cells and tissue. Then they induced the overexpression of miR-107 and performed functional tests to evaluate its effects. In the colony formation test, they found that the increase in the expression of miR-107 significantly inhibited cell proliferation. They complemented this finding with the flow cytometric cell cycle test and concluded that the inhibition resulted from an arrest in the G1/S phase of the cycle, due to the binding of miR-107 to cyclin E1. In the migration and invasion tests, they did not find any influence of miR-107 on these cellular processes [13].

Previous studies show the tumorigenicity of the miR-107 family [50–53]; thus, the inhibition of the expression of miR-107 might be a target for the treatment. miR-107 may promote the progression of prostate cancer to CRPCa, the end-stage of a multifactorial and heterogeneous disease process [54, 55]. Significant progress in understanding the molecular basis of CRPCa has been achieved in recent years [56] but despite this, CRPCa remains a lethal disease [57].

Recently, Liang et al. [58] reported that miR-107 induces chemoresistance in colorectal cancer (CRC) through the CAB39-AMPK-mTOR pathway, promoting metastasis. In this context, miR-107 levels could potentially be determined at the time of diagnosis to identify patients with aggressive disease/micro metastases and/or to predict recurrence following primary treatment.

Studies of Bryant et al. [48] strongly suggest that the presence of miR-107 in plasma can be used as a biomarker for cancer detection, monitoring, and prognosis predictor in PCa patients. They reported 12 miRNAs differentially expressed in the plasma of PCa patients compared to controls. Among these 12 miRNAs, 11 were significantly correlated with metastases. The association of two miRNAs, miR-141 and miR-375 with metastatic PCa was confirmed in a separate cohort. Furthermore, an analysis of urine revealed that miR-107 and miR-574-3p were also notably associated with PCa risk [48].

The role of miR-107 in other hallmarks of prostate cancer is not yet known. However, it would be relevant for the medical and scientific community to determine the participation of this miRNA in the development, growth, and metastasis of this disease; this will encourage further research tending to mitigate this pathology. The pleiotropic functions of miR-107 in diverse types of cancer indicate that it targets a variety of genes involved in tumor proliferation, invasiveness, metastasis, angiogenesis, and response to chemotherapy. Because of their carcinogenic or cancer-suppressor effects, miR107 can be used as a potential diagnostic and prognostic biomarker, or as a target for therapeutic intervention [48].

5. miR-107 and its participation in other urologic cancers

Both men and women can develop urological cancers, such as urethral, bladder, and kidney cancers. Other urologic cancers are gender specific. Males, besides the prostate, can experience testicular and penile cancer. The role of miR-107 in some of these cancers has not been reported previously.

Investigation of bladder cancer-related miRNAs shows a specific downregulation of miR-107 in the *in-situ* carcinoma lesions in comparison to a normal bladder [49]. Several studies have shown that miR-107 sponge effects could be involved in processes that upregulate circular RNAs (circRNA) in bladder cancer-related pathways. For example, overexpression of the circRNAs of the transcription factor (TCF) 25 (circ-TCF25) and transferrin receptor (TFRC) (circ-TFRC) negatively correlated with miR-107 promotes progression of bladder cancer through the circ-TCF25-miR-103a-3p/miR-107-CDK6 and circ-TFRC-miR-107-TFRC pathways, respectively [59, 60].

So far, only one study mentioning a relationship between miR-107 and kidney cancer has been published. Song et al. found that cell proliferation and invasiveness of renal cell carcinoma, which is the most common type of kidney cancer in adults, can be suppressed by high expression of miR-107 through an apparent cell cycle arrest at the G2/M phase [61]. In contrast, it has been found that high expression of miR-107 is frequently associated with a bad prognosis in patients with penile cancer [25]. Overexpression of miR-107 in penile cancer tissues was also reported by Zhang et al. [62].

6. Strategy and methodology of study of miR-107

While several techniques and methodologies have been used for the study of miR-107, the most used techniques are qRT-PCR, proliferation assays, Western blotting, luciferase reporter, and immunofluorescent and immunohistochemical assays (**Figure 1**, **Table 1**). Yet, other authors have used less conventional methodologies. Thus, the optimized electrochemical sensor technic enabled the PCR-free

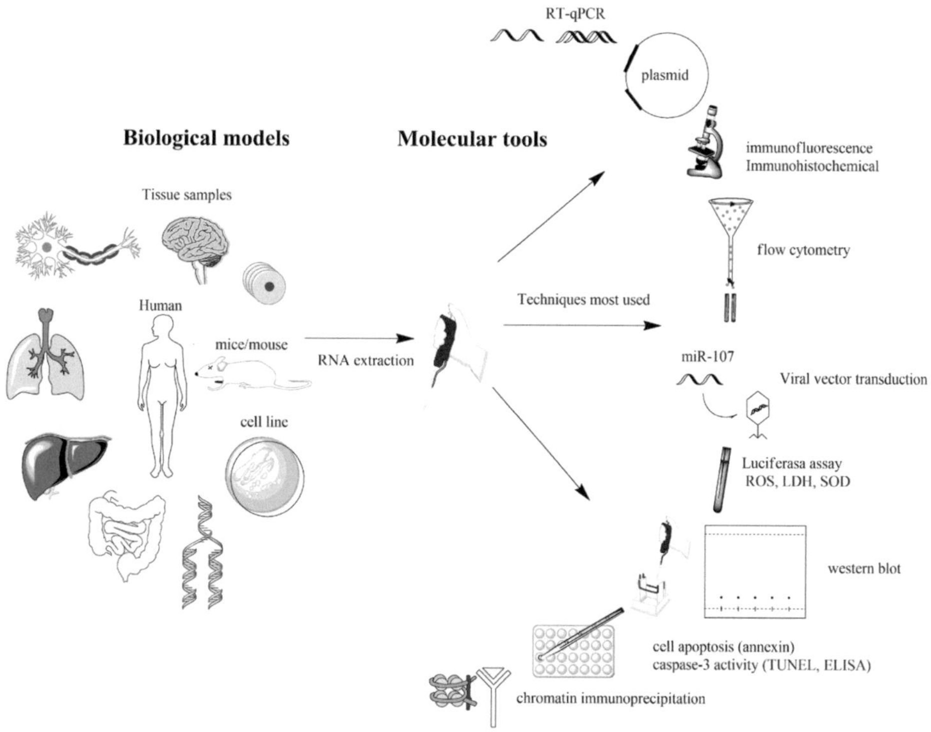

Figure 1.
Methodology most frequently used in the study of miR-107.

Pre-analysis	Techniques/molecular tools	Post-analysis		Authors
Organism/ pre-treatment	**Methodology**	**Statistical analysis**	**Conclusions**	
Mice (LDLr $^{-/-}$): fed a high-fat (liver tissue). *Hibiscus sabdarifa*: polyphenols. quercetin3-O-β-D-glucuronide	chromatography column, LC-MS, RT-PCR, Western blot, GC-MS, liver histology to measure the lipid droplet content.	ANOVA,U Mann–Whitney, t-Student's, Fisher's test $P< 0.05$	Polyphenols reduced the expression of miR-107 in the liver	[63]
Human glioma tissue. U251 and MO59K cell line. BALB/C-nu athymic nude mice.	qRT-PCR, transfection (miR-107), proliferation assay (MTT), Anchorage-independent growth assay, Cell apoptosis (annexing), luciferase reporter, Lentiviral construct transduction, Western blot and immunofluorescence analysis, computer-aided algorithms from PicTar.	Student's t-tests, Spearman's rank correlation. $P< 0.05$.	upregulation of miR-107 suppressed glioma cell growth	[64]
HepG2 cells and HEK 293.	cell transfection (miR-107), western blot, RT-PCR, DNA constructs, luciferase assay, Triglyceride assay,	student's t-test. $P<0.05$	miR-107 induced lipid accumulation in hepatocytes.	[65]
52 tumor of Kidney cancer. HKC cell line. Male nude mice	RT-PCR, Western blot, plasmid construction, transfection/ infection, luciferase reporter assay, flow cytometric	Wilcoxon rank sum test, ANOVA, Student t test.	miR-107 inhibit cell proliferation in renal cell carcinoma.	[61]
Primary cortical neurons, HEK293 and SHSY5Y cells. Osthole (7-methoxy-8-isopentenoxy coumarin)	Viral vector transduction, activity of LDH, cellular apoptosis (TUNEL), Immunofluorescence, RT-PCR, transfection (miR-107), western blot.	ANOVA, Bonferroni's, $P< 0.05$.	Osthole is neuroprotective via up-regulate miR-107 in AD.	[66]
SGC-7901, MKN-45, KATO III, BGC-823, AGS, MKN-28 and MKN1 cells lines. male nude mice: implanted miRNA-107 (tumor)	Transfection (miRNA-107), si-RNA, plasmid, RT-PCR, wound healing assay, luciferase activity, Western blot.	ANOVA, dunnett's multiple. $P< 0.05$	miR-107 acts as tumor inhibitor for gastric cancer	[67]
primary preadipocyte (mice)	Plasmid vector, transfection, cDNA synthesis, qPCR, luciferase reporter assays, caspase-3 activity, flow cytometry (apoptosis), TUNEL, western bloth, immunofluorescent.	Student's t-test. $P< 0.05$ and < 0.01.	miR-107 induced apoptosis pathway.	[68]
SGBS cells:	qPCR, western blotting, Transfected SGBS cells (miR-107, $[^{3}H]$palmitic acid).	not reported	miR-107 reduces adipogenesis	[69]

Pre-analysis	Techniques/molecular tools	Post-analysis		Authors
Organism/ pre-treatment	**Methodology**	**Statistical analysis**	**Conclusions**	
HCC patients: 1.FFPE cohort (tissue) 2. serum cohort (serum) Patient receiving before cisplatin, lipiodol, doxorubicin.	qPCR. Total RNA	Chi-square test, Mann-Whitney U. $P \leq 0.05$	miR-107 is biomarkers for predication of TACE treatment outcomes in HCC patients.	[70]
42 OA patients (cartilage samples): chondrocytes 48 rats: establish OA model	Transfection (chondrocytes/miR-107), flow cytometry, apoptosis (annexin/TUNEL assay), luciferase reporter, western blot, RT-q PCR,	Student's t-test. $P \leq 0.05$	miR-107 induced apoptosis and autophagy of OA chondrocytes	[71]
PC-3, DU145, LNCaP, 267B1, X/267B1, and Ki/267B1 cells	RNA isolation, qRT-PCR, electrochemical measurements (CHI 660D electrochemical workstation): frequency range of 0.1 Hz to 100 kHz with an alternating current amplitude of 10 mV.	Student's t-test. $P < 0.05$	Viability of the electrochemical evaluation method in clinical environment.	[72]
75 ADHD patients: stimulation in rDLPFC. for 6 weeks: 18 healthy children. Venous blood in both case.	RNA extraction, serum miRNA extraction, miRNA reverse transcription, fluorescence qPCR.	Mann-Whitney U test or Student's t-test. $P < 0.05$	serum miRNA-let-7d in ADHD patients is higher as compared to healthy children.	[73]
SKOV3 and 293T cells; transfected with lentiviruses	qRT-PCR, luciferase report assay, western blot, xenograft model and immunohistochemistry assay.	ANOVA, Student's t test, $P<0.05$	miR-107 as a tumor suppressor in ovarian cancer.	[74]
Patients with AD. SH-SY5Y and PC12 cells Mouse model: introducing 6-OHDA into the right ventral tegmental area	Cell transfection, qPCR, caspase-3 activity (ELISA), ROS, LDH, SOD, luciferase reporter assay, western blot. Rota-rod test. patients with AD: motor imagery test (fMRI)	ANOVA, Tukey's test, $P < 0.05$.	miR-107 may be a therapeutic target for the treatment of AD-related impairments.	[75]
HT-22 cell line	coimmunoprecipitation, chromatin immunoprecipitation, Luciferase reporter, qRT-PCR, apoptosis detection (annexin), western blot.	Student's t-test, ANOVA, Tukey's test $P < 0.05$	RMST activates p53/miR-107 signaling pathway	[76]
Serum samples (NSCLC). cell line BESA-2B, NSCLC, A549, H1299, HCC827, PC-9, 95-D, H1975, HEK-293T	luciferase reporter gene, qRT-PCR, western blot, Immunohistochemistry.	t-tests for two-group comparisons$P < 0.05$.	miR-107 could inhibit the progression of NSCLC	[77]

Pre-analysis	Techniques/molecular tools	Post-analysis		Authors
Organism/ pre-treatment	**Methodology**	**Statistical analysis**	**Conclusions**	
40 Sprague Dawley (SD) rats. MC3T3-E1 cells *Agrimonia pilos* (isolated polysaccharide) Rats induced SANFH (dex)	Femoral head tissue: apoptosis cellular (TUNEL), RT-PCR, cell proliferation (annexin), MC3T3-E1 cells transfected with anti-miR-107, flow cytometry, ALP activity, western blot.	ANOVA. $P< 0.05$	polysaccharides promote cell proliferation/ osteogenic differentiation by enhancing miR-107	[78]
HeLa and HEK	Reporter gene assay (FuGENE 6), RT-PCR, immunoblot/ Immunofluorescence assay, chromatin immunoprecipitation, ChIP-seq analysis: microRNA103–3p/107 target,. motif discovery (RSAT), Gene ontology analysis, genomic distribution (CEAS), siRNA transfections.	ANOVA, Bonferroni, KruskalWallis, Dunn's, t-test, Mann-Whitney U. $P< 0.05$.	miR-107 is potent regulators of GR function	[79]
30 HIBD rats Neonatal male Sprague-Dawley: HIBD injected with adenovirus: primary neuron cells	Neuronal cells (infected with adenovirus): RT-qPCR, Protein extraction, western blot, luciferase reporter, immunofluorescence, fluorescent hybridization, RNA immunoprecipitation. hippocampal neurons (NisslStaining).	unpaired t-test, ANOVA, Tukey test. $P < 0.05$.	miR-107 network has the potential to provide novel insights into treatment targets for HIBD.	[80]
70 tumor human tissue (NPC) Fresh frozen cell lines: HNE1, HONE1, 5–8F, 6–10B, and C666-1 and NP69:	qRT-PCR, western blot, cell apoptosis (anexin), flow cytometry, DIG-UTP-labeled miR-107, FISH kit, Immunofluorescence/ Immunohistochemical analysis BALB/c mice: (transfected HNE1 cells) lung	Student's t-test/Chi-square test, Pearson correlation. $P< 0.05$	circTGFBR2 suppresses the proliferation, migration, and invasiveness of NPC cells by sponging miR- 107	[81]

HIBD: hypoxic-ischemic brain damage. NPC: nasopharyngeal carcinoma. NP69: immortalized nasopharyngeal epithelial cell line. RT-qPCR: quantitative polymerase chain reaction. HeLa: Human cervical carcinoma cells. HEK: human embryonic kidney cells. GR: glucocorticoids. SANFH: steroid-induced osteonecrosis of the femoral head. Dex: dexamethasone. ALP: Alkaline phosphatase. BESA-2B, NSCLC, A549, H1299, HCC827, PC-9, 95-D, H1975: Human normal bronchial epithelial cell line: HEK-293T: human embryonic kidney cell line. NSCLC: Non-small cell lung cancer. TGFβR2: Transforming growth factor β receptor 2. HT-22: Immortalized mouse hippocampal neuron cell line. RMST: rhabdomyosarcoma 2-associated transcript. hnRNPK: Heterogeneous nuclear ribonucleoprotein K. SH-SY5Y: human neuroblastoma cell line. PC12: cells rat pheochromocytoma. LDH lactate dehydrogenase release. SOD: superoxide dismutase. ROS: reactive oxygen species. 6-OHDA: 6-hydroxydopamine. fMRI: magnetic resonance imaging. AD: Alzheimer's disease. ADHD: attention deficit hyperactivity disorder. rDLPFC: right dorsolateral prefrontal cortex. A549: human lung carcinoma cells. Human_Br07: human-origin seasonal influenza A virus H3N2. PC-3, DU145 and LNCaP: human transformed prostate epithelial cell lines. 267B1, X/267B1: cell line nontumorigenic in nude mice. Ki/267B1: cell line moderately tumorigenic nontumorigenic in nude mice. OA: Osteoarthritis. TRAF3: TNF Receptor Associated Factor 3. HCC: hepatocellular carcinoma. FFPE: Formalin-Fixed Paraffin-Embedded. TACE: transcatheter arterial chemoembolization (cancerous tumor therapy). SGBS: Simpson-Golabi-Behmel syndrome preadipocytes. LDLr -/-: deficient in LDL receptor. LC-MS: Liquid chromatography–mass spectrometry. GC-MS: Gas chromatography–mass spectrometry. U251 and MO59K: Human glioma cell lines. MTT: 3-(4,5-dimethylthiazol-2-yl)-2,5-diphenyltetrazolium bromide. SALL4: Sal-like 4. HKC: Human renal proximal tubular epithelial cell line. SHSY5Y: neuron cell line. SGC-7901, MKN-45, KATO III, BGC-823, AGS, MKN-28 and MKN1: gastric cancer cell lines. TGFBR2: transforming growth factor-β receptor II.

Table 1.
Methodology and techniques used in the study of miR-107.

quantification of miR-375 in CaP cells with acceptable specificity, confirming its potential applicability for point care (POC) purposes [72]. Detection at attomole levels of miRNA in samples is possible by electrocatalytic detection using gold-loaded nanoporous superparamagnetic iron oxide nanocubes, that has proved successful in the detection of miR-107 from cancer cell lines [82]. Other authors mentioned the use of chromatographic techniques such as Liquid chromatography-mass spectrometry (LC–MS) to extract and identify polyphenols from the plant *Hibiscus sabdarifa*, which reduced the expression of miR-107 in the liver [63] as well as of osthole (a coumarin compound) which is neuroprotective [66], and polysaccharides from *Agrimonia pilosa*, which promote cell proliferation enhanced by miR-107 [78, 83].

A variety of statistical tests have been used to validate the results, including parametric (ANOVA, student-*t*, Tukey) and non-parametric (Mann–Whitney U, Kruskal Wallis, Bonferroni, Dunn's multiple comparison, Chi-square, and Pearson correlation) tests. The probability level in all cases have been $P < 0.05$. In conclusion, the new technology and the use of diverse statistical tools validate the study and significance of mirR-107 in diverse biological situations, including CaP.

7. miR-107 as a possible blood biomarker

In 2019, Parvaee et al., evaluated the expression of three miRNAs, including miR-107, in blood samples from 50 patients with gastrointestinal cancer. After extracting and analyzing the plasma, they observed that the expression of miR-107 was significantly lower in patients with this type of cancer, compared to healthy volunteers (93.8% sensitivity and 78.8% specificity). From the ROC curve evaluation, they found that the patients could be distinguished from healthy people at cutoff levels of 0.504 (miR-107), 0.266 (miR-194), and 0.394 (miR-210). Finally, they concluded that miR-107 could serve as a possible plasma biomarker, assuring the minimum degree of invasion for the patient and an adequate level of reliability in the diagnosis [15].

8. The future for the miR-107

Given its role in cell cycle arrest and its direct involvement in cancer, the therapeutic potential of miR-107 is anticipated [13]. miR-107 may be a key target for the treatment of prostate cancer, arresting tumor growth, and cell survival.

9. Conclusion

The findings condensed in this review enable us to envisage miR-107 as a biomarker and possible therapeutic target for diverse types of prostate cancer. Although it is necessary to deepen the research on this molecule to explore its potential utility in the diagnosis, follow-up, and prognosis of cancer, our results support a role of miR-107 in the signaling cascades that allow cancer progression, and as shown here, in the progression of PCa to CRPCa. Our research suggests a potential utility of miR-107 as an accurate tool for diagnoses and follow-up of PCa.

Acknowledgements

We acknowledge the Universidad Autónoma de la Ciudad de México (UACM), the Colegio de Ciencia y Tecnología of the UACM (project number

PI-CCyT-2019-07), and the Secretaría de Posgrado e Investigación (SEPI) de la Escuela Nacional de Ciencias Biológicas del Instituto Politécnico Nacional, for their support in producing this article.

Funding information

The present study was funded by Universidad Autónoma de la Ciudad de México and the Consejo Nacional de Ciencia y Tecnología (CONACYT) (scholarship 734919 to EARC), and the Colegio de Ciencia y Tecnología of the UACM project PI-CCyT-2019-07 (19-09-4000.001-000-000-000-7100-0010).

Conflict of interest

The authors declare that they have no competing interests.

Author's contribution

MEAS, EARC, and EEP conceived, designed, and wrote the review. ORE, CLC, MDPR, JCTR, MCN and RAF reviewed and edited the manuscript. All authors read and approved the manuscript and agreed to be accountable for all aspects of the research ensuring that the accuracy or integrity of any part of the work is appropriately investigated and resolved.

Author details

Maria Elizbeth Alvarez-Sanchez[1*], Oscar Rojas Espinosa[2], Julio César Torres-Romero[3], Ereth Ameyatzin Robles Chávez[1], Edgar Estrella-Parra[4], María Dolores Ponce Regalado[5], Raúl Aragón Franco[6], Jose Gadú Campos Saucedo[7], Cesar López-Camarillo[1], Minerva Camacho Nuez[1] and Verónica Fernández Sánchez[4]

1 Posgrado en Ciencias Genómicas, Universidad Autónoma de la Ciudad de México, Ciudad de México, México

2 Departamento de Inmunología, Escuela Nacional de Ciencias Biológicas, Instituto Politécnico Nacional (IPN), Ciudad de México, México

3 Laboratorio de Bioquímica y Genética Molecular, Facultad de Química, Universidad Autónoma de Yucatán, Mérida, Yucatán, México

4 Facultad de Estudios Superiores Iztacala, Universidad Nacional Autónoma de México (UNAM), Ciudad de México, México

5 Departamento de Ciencias de la Salud, Centro Universitario de los Altos, Universidad de Guadalajara, Tepatitlán de Morelos, Jalisco, México

6 Universidad de la Salud "Vasco de Quiroga", Ciudad de México, México

7 Hospital Central Militar, Dirección General de Sanidad SEDENA, Ciudad de México, México

*Address all correspondence to: maria.alvarez@uacm.edu.mx

IntechOpen

© 2021 The Author(s). Licensee IntechOpen. This chapter is distributed under the terms of the Creative Commons Attribution License (http://creativecommons.org/licenses/by/3.0), which permits unrestricted use, distribution, and reproduction in any medium, provided the original work is properly cited.

References

[1] Hanahan D, Weinberg RA. The hallmarks of cancer. *Cell*. Published online 2000. doi:10.1016/S0092-8674(00)81683-9

[2] Hanahan D, Weinberg RA. Hallmarks of cancer: The next generation. *Cell*. Published online 2011. doi:10.1016/j.cell.2011.02.013

[3] The most common types of cancer [Internet]. 2021. Available from: https://gco.iarc.fr/today/home [Accessed: 2021-03-01]

[4] World Cancer Report 2014 [Internet]. 2020. Available from: https://www.who.int/cancer/publications/WRC_2014/en/ [Accessed: 2020-07-01]

[5] World Cancer Day: I Am and I Will [Internet]. 2021. Available from: https://www.paho.org/hq/index.php?option=com_content&view=article&id=15687:world-cancer-day-2020-i-am-and-i-will&Itemid=39809&lang=en [Accessed: 2021-03-01]

[6] Liu F, Liu S, Ai F, et al. MiR-107 promotes proliferation and inhibits apoptosis of colon cancer cells by targeting prostate apoptosis response-4 (Par4). Oncol Res. 2017;25(6):967-974. doi:10.3727/096504016X14803476672380

[7] Bartel DP. MicroRNAs: Genomics, Biogenesis, Mechanism, and Function. *Cell*. Published online 2004. doi:10.1016/S0092-8674(04)00045-5

[8] Lin S, Gregory RI. MicroRNA biogenesis pathways in cancer. Nat Rev Cancer. 2015;15(6):321-333. doi:10.1038/nrc3932

[9] Lu J, Getz G, Miska EA, et al. MicroRNA expression profiles classify human cancers. Nature. 2005;435(7043):834-838. doi:10.1038/nature03702

[10] Calin GA, Croce CM. MicroRNA signatures in human cancers. *Nat Rev Cancer*. Published online 2006. doi:10.1038/nrc1997

[11] Sita-Lumsden A, Dart DA, Waxman J, Bevan CL. Circulating microRNAs as potential new biomarkers for prostate cancer. *Br J Cancer*. Published online 2013. doi:10.1038/bjc.2013.192

[12] Norma Oficial Mexicana NOM-025-SSA3-2013, Para la organización y funcionamiento de las unidades de cuidados intensivos - 17 de Septiembre de 2013 - DOF. Diario Oficial de la Federación - Legislación - VLEX 461013722. Accessed December 2, 2020. https://dof.vlex.com.mx/vid/norma-nom-unidades-cuidados-intensivos-461013722

[13] Zhang X, Jin K, Luo JD, Liu B, Xie LP. MicroRNA-107 inhibits proliferation of prostate cancer cells by targeting cyclin E1. Neoplasma. 2019;66(5):704-716. doi:10.4149/neo_2018_181105N825

[14] Mitchell PS, Parkin RK, Kroh EM, et al. Circulating microRNAs as stable blood-based markers for cancer detection. *Proc Natl Acad Sci U S A*. Published online 2008. doi:10.1073/pnas.0804549105

[15] Parvaee P, Sarmadian H, Khansarinejad B, Amini M, Mondanizadeh M. Plasma level of microRNAs, miR-107, miR-194 and miR-210 as potential biomarkers for diagnosis intestinal-type gastric cancer in human. *Asian Pacific J Cancer Prev*. 2019;20(5):1421-1426. doi:10.31557/APJCP.2019.20.5.1421

[16] hsa-miR-107 targets [Internet]. 2020. Available from: http://www.mirdb.org/cgi-bin/search.cgi [Accessed: 2020-07-01]

[17] Lekchnov EA, Amelina E V., Bryzgunova OE, et al. Searching for the novel specific predictors of prostate cancer in urine: The analysis of 84 miRNA expression. *Int J Mol Sci*. Published online 2018. doi:10.3390/ijms19124088

[18] Yu QF, Liu P, Li ZY, et al. MiR-103/107 induces tumorigenicity in bladder cancer cell by suppressing PTEN. *Eur Rev Med Pharmacol Sci*. 2018;22(24):8616-8623. doi:10.26355/eurrev_201812_16625

[19] Zhou C, Li G, Zhou J, Han N, Liu Z, Yin J. Mir-107 activates ATR/chk1 pathway and suppress cervical cancer invasion by targeting MCL. PLoS One. 2014;9(11):111860. doi:10.1371/journal.pone.0111860

[20] Ai H, Zhou W, Wang Z, Qiong G, Chen Z, Deng S. microRNAs-107 inhibited autophagy, proliferation, and migration of breast cancer cells by targeting HMGB1. J Cell Biochem. 2019;120(5):8696-8705. doi:10.1002/jcb.28157

[21] Tang Z, Fang Y, Du R. MicroRNA-107 induces cell cycle arrests by directly targeting cyclin E1 in ovarian cancer. Biochem Biophys Res Commun. 2019;512(2):331-337. doi:10.1016/j.bbrc.2019.03.009

[22] Fu Y, Lin L, Xia L. MiR-107 function as a tumor suppressor gene in colorectal cancer by targeting transferrin receptor 1. *Cell Mol Biol Lett*. 2019;24(1). doi:10.1186/s11658-019-0155-z

[23] Wang L, Li K, Wang C, Shi X, Yang H. miR-107 regulates growth and metastasis of gastric cancer cells via activation of the PI3K-AKT signaling pathway by down-regulating FAT4. Cancer Med. 2019;8(11):5264-5273. doi:10.1002/cam4.2396

[24] Na C, Li X, Zhang J, Han L, Li Y, Zhang H. *MiR-107 Targets TRIAP1 to Regulate Oral Squamous Cell Carcinoma Proliferation and Migration*. Vol 12. e-Century Publishing Corporation; 2019.

[25] Pinho JD, Silva GEB, Teixeira Júnior AAL, et al. MIR-107, MIR-223-3P and MIR-21-5P Reveals Potential Biomarkers in Penile Cancer. *Asian Pac J Cancer Prev*. 2020;21(2):391-397. doi:10.31557/APJCP.2020.21.2.391

[26] Ciatto S, Zappa M, Bonardi R, Gervasi G. Prostate cancer screening: The problem of overdiagnosis and lessons to be learned from breast cancer screening. *Eur J Cancer*. Published online 2000. doi:10.1016/S0959-8049(00)00119-2

[27] Dall'era MA, Cooperberg MR, Chan JM, et al. Active surveillance for early-stage prostate cancer: Review of the current literature. *Cancer*. Published online 2008. doi:10.1002/cncr.23373

[28] Heinemann V, Boeck S, Hinke A, Labianca R, Louvet C. Meta-analysis of randomized trials: Evaluation of benefit from gemcitabine-based combination chemotherapy applied in advanced pancreatic cancer. *BMC Cancer*. Published online 2008. doi:10.1186/1471-2407-8-82

[29] Jemal A, Siegel R, Ward E, Hao Y, Xu J, Thun MJ. Cancer Statistics, 2009. *CA Cancer J Clin*. Published online 2009. doi:10.3322/caac.20006

[30] Sultana A, Tudur Smith C, Cunningham D, Starling N, Neoptolemos JP, Ghaneh P. Meta-analyses of chemotherapy for locally advanced and metastatic pancreatic cancer: Results of secondary end points analyses. *Br J Cancer*. Published online 2008. doi:10.1038/sj.bjc.6604436

[31] Loriot Y, Massard C, Fizazi K. Recent developments in treatments targeting castration-resistant prostate cancer bone metastases. *Ann Oncol*.

Published online 2012. doi:10.1093/annonc/mdr573

[32] Velonas VM, Woo HH, dos Remedios CG, Assinder SJ. Current status of biomarkers for prostate cancer. *Int J Mol Sci*. Published online 2013. doi:10.3390/ijms140611034

[33] Karrich JJ, Jachimowski LCM, Libouban M, et al. MicroRNA-146a regulates survival and maturation of human plasmacytoid dendritic cells. *Blood*. Published online 2013. doi:10.1182/blood-2012-12-475087

[34] Xu XH, Li DW, Feng H, Chen HM, Song YQ. MiR-300 regulate the malignancy of breast cancer by targeting p53. *Int J Clin Exp Med*. Published online 2015.

[35] Maebayashi T, Abe K, Aizawa T, et al. Solitary pulmonary metastasis from prostate cancer with neuroendocrine differentiation: A case report and review of relevant cases from the literature. *World J Surg Oncol*. Published online 2015. doi:10.1186/s12957-015-0598-2

[36] Baltimore D, Boldin MP, O'Connell RM, Rao DS, Taganov KD. MicroRNAs: New regulators of immune cell development and function. *Nat Immunol*. Published online 2008. doi:10.1038/ni.f.209

[37] Jerónimo C, Bastian PJ, Bjartell A, et al. Epigenetics in prostate cancer: Biologic and clinical relevance. *Eur Urol*. Published online 2011. doi:10.1016/j.eururo.2011.06.035

[38] Brase JC, Johannes M, Schlomm T, et al. Circulating miRNAs are correlated with tumor progression in prostate cancer. *Int J Cancer*. Published online 2011. doi:10.1002/ijc.25376

[39] He L, Hannon GJ. MicroRNAs: Small RNAs with a big role in gene regulation. *Nat Rev Genet*. Published online 2004. doi:10.1038/nrg1379

[40] Kosaka N, Iguchi H, Ochiya T. Circulating microRNA in body fluid: A new potential biomarker for cancer diagnosis and prognosis. *Cancer Sci*. Published online 2010. doi:10.1111/j.1349-7006.2010.01650.x

[41] Hasselmann DO, Rappl G, Tilgen W, Reinhold U. Extracellular tyrosinase mRNA within apoptotic bodies is protected from degradation in human serum. *Clin Chem*. Published online 2001. doi:10.1093/clinchem/47.8.1488

[42] Cocucci E, Racchetti G, Meldolesi J. Shedding microvesicles: Artefacts no more. Trends Cell Biol. 2009;19(2): 43-51. doi:10.1016/j.tcb.2008.11.003

[43] Tsujiura M, Komatsu S, Ichikawa D, et al. Circulating miR-18a in plasma contributes to cancer detection and monitoring in patients with gastric cancer. Gastric Cancer. 2015;18(2): 271-279. doi:10.1007/s10120-014-0363-1

[44] Kawaguchi T, Komatsu S, Ichikawa D, et al. Circulating microRNAS: A next-generation clinical biomarker for digestive system cancers. *Int J Mol Sci*. Published online 2016. doi:10.3390/ijms17091459

[45] Komatsu S, Ichikawa D, Kawaguchi T, et al. Plasma microRNA profiles: Identification of miR-23a as a novel biomarker for chemoresistance in esophageal squamous cell carcinoma. *Oncotarget*. Published online 2016. doi:10.18632/oncotarget.11500

[46] Reinhart BJ, Slack FJ, Basson M, et al. The 21-nucleotide let-7 RNA regulates developmental timing in *Caenorhabditis elegans*. *Nature*. Published online 2000. doi:10.1038/35002607

[47] Michael MZ, O'Connor SM, Van Holst Pellekaan NG, Young GP, James RJ.

Reduced Accumulation of Specific MicroRNAs in Colorectal Neoplasia. *Mol Cancer Res*. Published online 2003.

[48] Bryant RJ, Pawlowski T, Catto JWF, et al. Changes in circulating microRNA levels associated with prostate cancer. *Br J Cancer*. Published online 2012. doi:10.1038/bjc.2011.595

[49] Catto JWF, Miah S, Owen HC, et al. Distinct MicroRNA alterations characterize high- and low-grade bladder cancer. *Cancer Res*. Published online 2009. doi:10.1158/0008-5472.CAN-09-0744

[50] Chen H-Y, Lin Y-M, Chung H-C, et al. miR-103/107 Promote Metastasis of Colorectal Cancer by Targeting the Metastasis Suppressors DAPK and KLF4. *Cancer Res*. 2012;72(14):3631 LP - 3641. doi:10.1158/0008-5472.CAN-12-0667

[51] Zheng J, Liu Y, Qiao Y, Zhang L, Lu S. miR-103 Promotes Proliferation and Metastasis by Targeting KLF4 in Gastric Cancer. *Int J Mol Sci*. 2017;18(5):910. doi:10.3390/ijms18050910

[52] Xiong B, Lei X, Zhang L, Fu J. miR-103 regulates triple negative breast cancer cells migration and invasion through targeting olfactomedin 4. *Biomed Pharmacother*. 2017;89:1401-1408. doi: https://doi.org/10.1016/j.biopha.2017.02.028

[53] Xiong J, Wang D, Wei A, et al. Deregulated expression of miR-107 inhibits metastasis of PDAC through inhibition PI3K/Akt signaling via caveolin-1 and PTEN. Exp Cell Res. 2017;361(2):316-323. doi: https://doi.org/10.1016/j.yexcr.2017.10.033

[54] deVere White RW, Vinall RL, Tepper CG, Shi X-B. MicroRNAs and their potential for translation in prostate cancer. Urol Oncol Semin Orig Investig. 2009;27(3):307-311. doi: https://doi.org/10.1016/j.urolonc.2009.01.004

[55] Shi X-B, Xue L, Yang J, et al. An androgen-regulated miRNA suppresses Bak1 expression and induces androgen-independent growth of prostate cancer cells. *Proc Natl Acad Sci*. 2007;104(50):19983 LP - 19988. doi:10.1073/pnas.0706641104

[56] Lin SL, Chiang A, Chang D, Ying SY. Loss of mir-146a function in hormone-refractory prostate cancer. *RNA*. Published online 2008. doi:10.1261/rna.874808

[57] Sun T, Wang Q, Balk S, Brown M, Lee GSM, Kantoff P. The role of microrna-221 and microrna-222 in Androgen- independent prostate cancer cell lines. *Cancer Res*. Published online 2009. doi:10.1158/0008-5472.CAN-08-4112

[58] Liang Y, Zhu D, Hou L, et al. MiR-107 confers chemoresistance to colorectal cancer by targeting calcium-binding protein 39. *Br J Cancer*. Published online 2020. doi:10.1038/s41416-019-0703-3

[59] Zhong Z, Lv M, Chen J. Screening differential circular RNA expression profiles reveals the regulatory role of circTCF25-miR-103a-3p/miR-107-CDK6 pathway in bladder carcinoma. Sci Rep. 2016;6. doi:10.1038/srep30919

[60] Su H, Tao T, Yang Z, et al. Circular RNA cTFRC acts as the sponge of MicroRNA-107 to promote bladder carcinoma progression. Mol Cancer. 2019;18(1):27. doi:10.1186/s12943-019-0951-0

[61] Song N, Ma X, Li H, et al. MicroRNA-107 functions as a candidate tumor suppressor gene in renal clear cell carcinoma involving multiple genes. *Urol Oncol Semin Orig Investig*. Published online 2015. doi:10.1016/j.urolonc.2015.02.003

[62] Zhang L, Wei P, Shen X, et al. MicroRNA expression profile in penile

cancer revealed by next-generation small RNA sequencing. *PLoS One*. Published online 2015. doi:10.1371/journal.pone.0131336

[63] Joven J, Espinel E, Rull A, et al. Plant-derived polyphenols regulate expression of miRNA paralogs miR-103/107 and miR-122 and prevent diet-induced fatty liver disease in hyperlipidemic mice. Biochim Biophys Acta - Gen Subj. 2012;1820(7):894-899. doi:10.1016/j.bbagen.2012.03.020

[64] He J, Zhang W, Zhou Q, et al. Low-expression of microRNA-107 inhibits cell apoptosis in glioma by upregulation of SALL4. *Int J Biochem Cell Biol*. Published online 2013. doi:10.1016/j.biocel.2013.06.008

[65] Bhatia H, Verma G, Datta M. MiR-107 orchestrates ER stress induction and lipid accumulation by post-transcriptional regulation of fatty acid synthase in hepatocytes. *Biochim Biophys Acta - Gene Regul Mech*. Published online 2014. doi:10.1016/j.bbagrm.2014.02.009

[66] Jiao Y, Kong L, Yao Y, et al. Osthole decreases beta amyloid levels through up-regulation of miR-107 in Alzheimer's disease. *Neuropharmacology*. Published online 2016. doi:10.1016/j.neuropharm.2016.04.046

[67] Cheng F, Yang Z, Huang F, Yin L, Yan G, Gong G. microRNA-107 inhibits gastric cancer cell proliferation and metastasis by targeting PI3K/AKT pathway. Microb Pathog. 2018;121:110-114. doi:10.1016/j.micpath.2018.04.060

[68] Zhang Z, Wu S, Muhammad S, Ren Q, Sun C. miR-103/107 promote ER stress-mediated apoptosis via targeting the Wnt3a/−catenin/ATF6 pathway in preadipocytes. J Lipid Res. 2018;59(5):843-853. doi:10.1194/jlr.M082602

[69] Ahonen MA, Haridas PAN, Mysore R, Wabitsch M, Fischer-Posovszky P, Olkkonen VM. miR-107 inhibits CDK6 expression, differentiation, and lipid storage in human adipocytes. *Mol Cell Endocrinol*. Published online 2019. doi:10.1016/j.mce.2018.09.007

[70] Ali HEA, Emam AA, Zeeneldin AA, et al. Circulating miR-26a, miR-106b, miR-107 and miR-133b stratify hepatocellular carcinoma patients according to their response to transarterial chemoembolization. *Clin Biochem*. Published online 2019. doi:10.1016/j.clinbiochem.2019.01.002

[71] Zhao X, Li H, Wang L. MicroRNA-107 regulates autophagy and apoptosis of osteoarthritis chondrocytes by targeting TRAF3. *Int Immunopharmacol*. Published online 2019. doi:10.1016/j.intimp.2019.03.005

[72] Jeong B, Kim YJ, Jeong JY, Kim YJ. Label-free electrochemical quantification of microRNA-375 in prostate cancer cells. *J Electroanal Chem*. Published online 2019. doi:10.1016/j.jelechem.2019.05.009

[73] Cao P, Wang L, Cheng Q, et al. Changes in serum miRNA-let-7 level in children with attention deficit hyperactivity disorder treated by repetitive transcranial magnetic stimulation or atomoxetine: An exploratory trial. *Psychiatry Res*. Published online 2019. doi:10.1016/j.psychres.2019.02.037

[74] Tang Z, Fang Y, Du R. MicroRNA-107 induces cell cycle arrests by directly targeting cyclin E1 in ovarian cancer. *Biochem Biophys Res Commun*. Published online 2019. doi:10.1016/j.bbrc.2019.03.009

[75] Sun L, Zhang T, Xiu W, et al. MiR-107 overexpression attenuates neurotoxicity induced by 6-hydroxydopamine both in vitro and in vivo. *Chem Biol Interact*. Published

online 2020. doi:10.1016/j.cbi.2019.108908

[76] Cheng H, Sun M, Wang ZL, et al. LncRNA RMST-mediated miR-107 transcription promotes OGD-induced neuronal apoptosis via interacting with hnRNPK. *Neurochem Int*. Published online 2020. doi:10.1016/j.neuint.2019.104644

[77] Wu Z, Yuan Q, Yang C, et al. Downregulation of oncogenic gene TGFβR2 by miRNA-107 suppresses non-small cell lung cancer. *Pathol Res Pract*. 2020;216(1). doi:10.1016/j.prp.2019.152690

[78] Huang W, Jin S, Yang W, et al. Agrimonia pilosa polysaccharide and its sulfate derives facilitate cell proliferation and osteogenic differentiation of MC3T3-E1 cells by targeting miR-107. *Int J Biol Macromol*. Published online 2020. doi:10.1016/j.ijbiomac.2019.11.213

[79] Yang N, Berry A, Sauer C, et al. Hypoxia regulates GR function through multiple mechanisms involving microRNAs 103 and 107. *Mol Cell Endocrinol*. Published online 2020. doi:10.1016/j.mce.2020.111007

[80] Fang H, Li H-F, Pan Q, et al. Long noncoding RNA H19 overexpression protects against hypoxic-ischemic brain damage by inhibiting miR-107 and up-regulating vascular endothelial growth factor. Am J Pathol. 2021;191(3):503-514. doi:10.1016/j.ajpath.2020.11.014

[81] Li W, Lu H, Wang H, et al. Circular RNA TGFBR2 acts as a ceRNA to suppress nasopharyngeal carcinoma progression by sponging miR-107. *Cancer Lett*. Published online 2021. doi:10.1016/j.canlet.2020.11.001

[82] Islam MN, Masud MK, Nguyen NT, et al. Gold-loaded nanoporous ferric oxide nanocubes for electrocatalytic detection of microRNA at attomolar level. *Biosens Bioelectron*. Published online 2018. doi:10.1016/j.bios.2017.09.027

[83] Huang W, Deng H, Jin S, Ma X, Zha K, Xie M. The isolation, structural characterization and anti-osteosarcoma activity of a water soluble polysaccharide from Agrimonia pilosa. *Carbohydr Polym*. Published online 2018. doi:10.1016/j.carbpol.2018.01.047